Instructor's Manual

Technical Writing
Process and Product

Fourth Edition

Steven M. Gerson
Johnson County Community College

Sharon J. Gerson
DeVry Institute of Technology

Prentice Hall, Upper Saddle River, New Jersey 07458

Prentice Hall

Upper Saddle River, New Jersey 07458

10 9 8 7 6 5 4 3 2 1

ISBN 0-13-099324-7

Printed in the United States of America

TABLE OF CONTENTS

PREFACE

PREFACE

Introducing the Textbook

This Instructor's Manual provides a step-by-step approach to using *Technical Writing: Process and Product.* We include

- overviews of each chapter
- sample course outlines for 17-week semesters and 15-week semesters
- teaching suggestions
- possible answers to end-of-chapter activities
- answers to questions about clarity, conciseness, mechanics and grammar
- additional examples

Our unique approach to technical writing makes *Technical Writing: Process and Product* a very useful text. Usually, technical writing books briefly mention the writing process in a chapter early in the text. They rarely mention process again. Our text, in contrast, makes the writing process an integral part of technical writing.

Each chapter shows you how to use the writing process to create a type of technical writing (memos, letters, descriptions, instructions, etc.). We also include process logs, which are samples of student writing to reveal the process in action. These process logs start with prewriting, show a rough draft, provide sample revisions, and conclude with a finished copy. Thus, the process logs model the writing process for your students, providing them a writing technique they can emulate when constructing their own correspondence.

We have also included team writing activities and case studies at the end of most chapters. These case studies and team writing projects encourage your students to write collaboratively, interact with each other in small groups, and create work-related correspondence drawn from business-industry scenarios.

You will find that both you and your students will benefit from our practical approach to writing.

Understanding the Units

Unit 1—Defining Technical Writing.

Chapters 1-4 lay the foundation for the entire textbook. Chapter 1 (An Introduction to Technical Writing) provides an in-depth discussion of technical writing, distinguishing it from all other forms of discourse and explaining why technical writing is essential in today's world. We also discuss teamwork and collaborative writing techniques. This is an important chapter for students whose prior writing experiences might have been limited to expressive or expository essays. To such students, writing equals essays--and nothing else. If they come to technical writing with the same expectations they've held regarding essay construction, they could be confused by their new responsibilities. Our first chapter helps orient students to technical writing as a different type of written communication.

Chapter 2 (Producing the Product) explains the writing process, emphasizes usability testing (***New*** to this edition), and provides a sample process log to illustrate how the writing process works. Chapter 3 (Objectives in Technical Writing) discusses technical writing style and the importance of ethics when communicating for business and industry. ***New*** to this 4th edition, we focus on Internet ethics and online intellectual property laws. Chapter 4 (Audience Recognition and Involvement) discusses audience and the role of multiculturalism in a global economy.

We usually spend the first several weeks of each new semester teaching these four chapters. Doing so accomplishes two important goals: we clarify what successful technical writing entails, and we provide the students our criteria for assessing their work.

Unit 2—Correspondence.

Chapter 5 (Memos) and Chapter 6 (Letters) help your students write effective correspondence. To do so, we take you and your students through the writing process by

- providing different prewriting techniques (reporter's questions for letters; clustering for memos)
- explaining how to write rough drafts
- suggesting ways to revise
- including process logs complete with prewriting, rough drafts, peer evaluations, revisions, and a finished product

New to our 4th edition, we discuss the pros and cons of memo and letter Wizards.

Chapter 7 (The Job Search) explains various techniques used to get a job. We emphasize use of traditional search methods, such as newspapers, career centers, and networking. We also discuss online resources. In addition, writers are shown how to write effective letters of application, resumes, online resumes, and follow-up letters. ***New*** to this edition, we focus on three different types of resumes: mail versions, scannable versions, and Internet resumes.

Unit 3—Visual Appeal.

Chapters 8 and 9 focus on page layout and graphics. Chapter 8 (Document Design) shows how to achieve readability and access through highlighting, headings, space, typeface, type style, and numerous other techniques. Chapter 9 (Graphics) distinguishes tables from figures and provides criteria for constructing effective graphics. ***New*** to this 4th edition, we've included 4-color samples and new types of graphics, including combination charts, 3-D topographical charts, geologic maps, and CAD drawings.

Unit 4—Technical Applications.

Chapter 10 (***completely New for this edition***) focuses on writing fliers, brochures, and newsletters. We emphasize the uses of these types of correspondence, as well as organizational techniques, content, design elements, and usability testing.

Chapter 11 (Technical Descriptions) focuses on mechanism descriptions, including how to organize and develop descriptions using denotative words and graphics.

Chapter 12 (Instructions and User Manuals) is significantly revised for this edition. It discusses criteria for writing short instructions, including attention to organization, audience, graphics, hazard information, and style. In addition, we focus on longer user manuals, and include ***New*** examples with screen captures as well as revised features for usability testing.

Unit 5—Electronic Communication.

Chapter 13 (Electronic Communication) updates our discussion of e-mail, Web sites, and online help screens. To help students visualize this type of technical writing, we include screen captures of e-mail and inserts of a Web site home page and online help screen. This chapter also includes detailed information about HTML (hypertext markup language). Most importantly, in the ever-changing realm of electronic communication, this chapter includes ***New*** and updated information on e-mail netiquette, copyright concerns, Internet legalities, and usability considerations for web design.

Unit 6—Report Strategies.

Chapter 14 (Research) gives the reader techniques for researching information for reports. In addition to traditional forms of research, ***New*** this edition, we discuss online research techniques and sources, including Internet search engines, directories, and metasearch engines. We also discuss MLA and APA documentation, including methods for documenting Web-related sources.

In Chapter 15 (The Summary), we show the reader how to summarize a document. Chapter 16 (Reports) focuses on criteria for writing reports no longer than 10 pages. These include trip reports, progress reports, lab reports, feasibility reports, and incident reports. Chapter 17 (Proposals) uses the longer proposal to show readers how to write reports of more than 10 pages. Here we add specific criteria for these longer documents, including title pages, tables of content, abstracts, and appendices.

Chapter 18 (Oral Presentations) assists the reader with organization and delivery for impromptu, memorized, manuscript, and extemporaneous speeches. The reader will also find information about incorporating visual aids in the presentation.

Unit 7—Handbook.

Chapter 19 (Grammar, Punctuation, Mechanics, and Spelling) discusses the most common grammatical, punctuation, and mechanical errors found in technical documents. In addition, ***New*** to this edition, we include grammar quizzes.

Organizing Your Class

Teaching technical writing using the writing process (prewriting, writing, and rewriting) helps students write effectively. In addition to helping students, the writing process also is teachable. The overall organization of *Technical Writing: Process and Product* provides you with an effective teaching methodology. A process-approach to writing can help you minimize lecture and maximize your students' hands-on application.

Every chapter of our textbook gives you the following:

- chapter outlines
- objectives for each type of technical document
- criteria to follow
- unique prewriting techniques geared toward each type of technical writing
- organizational techniques for rough drafting
- guidelines for rewriting
- checklists to help readers make sure they have accomplished the goals of each writing activity
- process logs (samples of student-written correspondence which help your students emulate the writing process)
- samples from corporate communication and student writing for classroom discussion
- end-of-chapter activities, allowing individuals and teams to practice the different types of communication taught throughout the textbook
- collaborative writing projects and case studies
- chapter highlights
- chapter quiz questions

Using the above information, you can organize your course any way you choose. Each chapter in *Technical Writing: Process and Product* can be taught in any sequence. However, to give you some suggestions for course organizations using our text, following are sample course outlines for a 17-week semester and for a 15-week semester.

17- Week Semester Course Outline

Week 1 Introduction to course objectives and assignments
Chapters 1-2, define technical writing and explain the writing process

Week 2 Chapters 3-4, discuss clarity, conciseness, accuracy, organization, ethics, and audience. End-of-chapter activities

Week 3 Chapter 8, discuss document design. End-of-chapter activities
Chapter 6, discuss letters (essential components)
Discuss sales letters--objectives, criteria, and samples
Collaborative sales letter writing activity (refer to Chapter 1 for team-writing skills)--prewriting, writing a rough draft, rewriting through peer revision
Homework: prewrite, write rough draft of individual sales letter

Week 4 Collaborative sales letter reviewed and revisions discussed. Homework due. Peer evaluations of individual sales letters. Revisions based on peer group evaluations

Week 5 **Sales Letters due**
Chapter 6, discuss letter of inquiry--objectives, criteria, and samples
Collaborative letter of inquiry writing activity--prewriting, writing a rough draft,
rewriting through peer revision
Collaborative letter of inquiry reviewed and revisions discussed
Homework: prewrite, write rough draft of individual letter of inquiry

Week 6 Homework due. Peer evaluations of individual letter of inquiry
Revisions based on peer group evaluations
Letter of Inquiry due
Chapter 5, discuss memos--objectives, criteria, samples (see also Chapter 10-compare/contrast memos to e-mail)
Collaborative memo writing activity--prewriting, writing a rough draft, rewriting through peer revision
Homework: prewrite, write rough draft of individual memo

Week 7 Collaborative memos reviewed and revisions discussed
Homework due. Peer evaluations of individual memo
Revisions based on peer group evaluations
Memo due
Chapter 9, discuss graphics. End-of-chapter activities.

Week 8 Chapter 11, discuss technical descriptions--objectives, criteria, samples
Collaborative description writing activity--prewriting, writing a rough draft, rewriting through peer revision
Collaborative description reviewed and revisions discussed
Homework: prewrite, write rough draft of individual description

Week 9 Homework due. Peer evaluations of description
Revisions based on peer group evaluations
Technical Description due
Chapter 12, discuss instructions--objectives, criteria, and samples
Collaborative instruction writing activity--prewriting, writing a rough draft, rewriting through peer revision

Week 10 Collaborative instruction reviewed and revisions discussed
Homework: prewrite, write rough draft of instructions
Homework due. Peer evaluations of instruction
Revisions based on peer group evaluations

Week 11 **Technical Instructions due**
Chapter 16, discuss progress report--objectives, criteria, and samples
Homework: prewrite and write rough draft progress report
Homework due. Peer evaluations
Revisions based on peer group evaluations

Week 12 **Progress reports due**
Chapter 17, discuss proposals--objectives, criteria, and samples
Homework: prewrite
Homework due--tutorial review of prewriting
Chapter 14, research--objectives, criteria, and samples
Discussion of documentation and use of quoting and paraphrasing
Homework: rough drafts for abstract and introduction

Week 13 Homework due--rough drafts for peer evaluations
Revisions from peer group evaluations
Homework: rough drafts for proposal body
Homework due--peer evaluations of proposal body
Revisions from peer group evaluations
Homework: rough draft for proposal conclusion

Week 14 Homework due--rough drafts for peer evaluation
Revision from peer group evaluations
Chapter 6, cover letter--objectives, criteria, and samples
Homework: rough draft cover letter
Homework due--peer evaluations
Revisions from peer group evaluations

Week 15 **Proposal and Cover Letter due**
Chapter 7, resumes and letters of application--objectives, criteria, and samples
Homework: rough draft resume
Homework due--peer evaluations
Revision from peer group evaluations
Homework: rough draft letter of application

Week 16 Homework due--rough draft for peer evaluations
Revisions from peer group evaluations
Resume and Letter of Application due
Semester review for final exam

Week 17 Final Exam

15-Week Semester Course Outline

Week 1 Introduction to course objectives and assignments
Chapters 1-2, define technical writing and explain the writing process

Week 2 Chapters 3-4, discuss clarity, conciseness, accuracy, organization, ethics, and audience. End-of-chapter activities

Week 3 Chapter 8, discuss document design. End-of-chapter activities
Chapter 6, discuss letters (essential components)
Discuss letter of inquiry--objectives, criteria, and samples
Collaborative letter of inquiry writing activity (refer to Chapter 1 for team-writing skills)--prewriting, writing a rough draft, rewriting through peer revision
Collaborative letters reviewed and revisions discussed
Homework: prewrite, write rough draft of individual letter of inquiry

Week 4 Homework due. Peer evaluations of individual letters of inquiry
Revisions based on peer group evaluations
Letters of Inquiry due
Chapter 5, discuss memos--objectives, criteria, samples (see also Chapter 10- compare/contrast memos to e-mail)
Collaborative memo writing activity--prewriting, writing a rough draft, rewriting through peer revision
Homework: prewrite, write rough draft of individual memo

Week 5 Collaborative memos reviewed and revisions discussed
Homework due. Peer evaluations of individual memo
Revisions based on peer group evaluations
Memo due
Chapter 9. Discuss graphics. End-of-chapter activities.

Week 6 Chapter 11, discuss technical descriptions--objectives, criteria, samples
Collaborative description writing activity--prewriting, writing a rough draft, rewriting through peer revision
Collaborative description reviewed and revisions discussed
Homework: prewrite, write rough draft of individual description

Week 7 Homework due. Peer evaluations of description revisions based on peer group evaluations
Technical Description due
Chapter 12, discuss instructions--objectives, criteria, and samples
Collaborative instruction writing activity--prewriting, writing a rough draft, rewriting through peer revision

Week 8 Collaborative instruction reviewed and revisions discussed
Homework: prewrite, write rough draft of individual letter of inquiry
Homework due. Peer evaluations of instruction
Revisions based on peer group evaluations

Week 9 **Technical Instructions due**
Chapter 17, proposals--objectives, criteria, and samples
Homework: prewrite
Homework due--tutorial review of prewriting
Chapter 14, research--objectives, criteria, and samples

Week 10 Discussion of documentation and use of quoting and paraphrasing
In-class activities (documenting, quoting, and paraphrasing)
Homework: rough drafts for abstract and introduction

Week 11 Homework due--rough drafts for peer evaluations
Revisions from peer group evaluations
Homework: rough drafts for proposal body
Homework due--peer evaluations of proposal body
Revisions from peer group evaluations
Homework: rough draft for proposal conclusion

Week 12 Homework due--rough drafts for peer evaluation
Revision from peer group evaluations
Chapter 6, cover letter--objectives, criteria, and samples
Homework: rough draft cover letter
Homework due--peer evaluations
Revisions from peer group evaluations

Week 13 **Proposal and Cover Letter due**
Chapter 7, resumes and letters of application--objectives, criteria, and samples
Homework: rough draft resume
Homework due--peer evaluations
Revision from peer group evaluations
Homework: rough draft letter of application

Week 14 Homework due--rough draft for peer evaluations
Revisions from peer group evaluations
Resume and Letter of Application due
Semester review for final exam

Week 15 Final Exam

> **Note:** In the above sample syllabi, we haven't included all possible assignments or combinations of assignments. For example, you might want to replace the sales letter with a flier, the proposal with a web site or oral presentation, or the technical description with a brochure or newsletter.

As you noticed by reading these sample course outlines, much time is spent in class working with collaborative writing, prewriting, rough drafts, and peer group revisions. These are primary objectives of our text. *Technical Writing: Process and Product* guides the students toward their finished products through the process approach to writing. Not only does this process approach help the students construct effective technical correspondence, but also it accomplishes two additional goals.

1. The collaborative writing encourages teamwork, a total quality management ideal (see Chapter 1 for further information on this topic). We constantly use collaborative writing activities. Student team projects help involve everyone actively in writing. In a team project, no one student can sit passively, especially if his or her grade is tied to the writing activity. Collaboration also allows students to practice their technical writing skills and receive immediate feedback from their peers.
2. The process approach turns your class into a writing workshop where students spend time writing, reading, and revising rather than listening to lectures. The process approach minimizes teacher-focused instruction and maximizes your students' hands-on application.

To help you implement the above goals, we have provided many end-of-chapter activities. They include assignments that can be done individually or in teams. We also include case studies. Furthermore, the easily accessible criteria checklists in each chapter can be used for peer group evaluations. These checklists will help your students provide each other constructive criticism.

New to this 4th edition you will find two additional valuable resources:

- End-of-chapter quiz questions and answers to these questions
- Unit Exams (with questions and answers).

UNIT 1

DEFINING TECHNICAL WRITING

CHAPTER 1

AN INTRODUCTION TO TECHNICAL WRITING

@ Distinctions

@ Purposes

@ Importance

@ Teamwork

ABSTRACT

Teachers often are asked, "What is technical writing, and how does it differ from written communication taught in traditional composition classes?" To answer those questions, this chapter defines technical writing, differentiates it from other types of writing, and stresses the importance of technical writing in business and industry.

@ The Differences between Technical Writing and Other Types of Writing

We are all familiar with different types of writing, such as creative writing, expressive essays, expository writing, journalism, and technical writing. Our students, in contrast, are not aware of distinctions among these types of communication. *Technical Writing: Process and Product* explains how these types of writing differ.

However, to make the distinctions clearer to your students, you might want to elaborate on our discussion. Consider these possibilities:

1. ***Create an analogy***. We ask students to list different types of shoes they own, such as boots, loafers, tennis shoes, bedroom slippers, and dress shoes. Then we ask them why they own such diverse shoes. The answer is obvious; each type of shoe serves a unique purpose. They would not wear their dress shoes to shovel snow nor their tennis shoes to a wedding. The same holds true for writing. One type of writing does not fit all purposes. No one would write poetry, essays, or diary entries at work.

2. ***Chart the differences***. To further explain the analogy above, we provide the students samples of different types of writing, such as a poem, a journalistic editorial, an expressive essay, a persuasive movie review, and a technical report.

Poem	Journalism	Essay	Movie Review	Report

We ask the students to review these different types of writing and compare/contrast them on the basis of word *usage, sentence structure, page layout, and audience* (for example). The students then list the unique traits of each type of writing. This visual helps students clearly see how technical writing differs from other types of writing.

NOTE: If the above five different types of writing are too complex and too time consuming, limit the comparison/contrast to a typical, student-written essay and a technical report. Even that would be sufficient to clarify the difference between technical writing and other types of written communication.

@ The Importance of Technical Writing

You probably have encountered people in business who have suffered due to their poor written and verbal skills. You've talked to people who were not promoted because their writing was flawed, or you've heard stories about bosses rejecting reports because of misspelled words. Similarly, you have received flawed correspondence from companies and concluded that such companies were either unprofessional or incompetent. In contrast, you have talked to employees who were praised and respected for their writing skills--they've earned promotions or been given more responsibilities due to their writing abilities.

We clarify to our students how important technical writing is by sharing anecdotes. Nothing drives a point home more effectively than true stories. Share your personal experience stories with your students.

Another way to emphasize the importance of technical writing is by drawing on the persuasive powers of corporate spokespersons. Your students soon will be seeking employment, and companies in your community want to hire employees who can communicate successfully. These corporations know that the skills you're teaching are important. Invite business employees, personnel directors, and human resource supervisors to speak to your classes. Let the community work for you to prove that technical writing is a valuable skill. You could videotape their talks and then use these videotapes for future classes.

@ Teamwork

Total Quality Management (TQM) and Continuous Quality Improvement (CQI) are business practices that use teams of employees to solve corporate problems and to ensure quality production. To simulate the use of teamwork in the work place, we emphasize teamwork in the classroom. Our goal is to prepare students for their future careers.

You can help your students by assigning numerous team writing activities and/or team projects. These could include simple assignments, such as team-written letters or memos that take approximately one or two class periods. Longer team projects, such as team-written proposals or instructional manuals, take approximately two to four weeks.

Team writing is not easy. Students miss class, have different levels of ability, and varying degrees of enthusiasm. Good students are frustrated by weaker students; poor students coast on the abilities of better students. Such challenges are inevitable, but they're worth the trouble. Solving the problems that necessarily accompany teamwork also helps prepare your students for their future careers.

We try to reward the good team players and punish the slackers through our grading procedures. For example, if a team project (such as a long, instructional manual) is worth 25 points, we break the grade down as follows:

- 15 points = the student's designated part of the team project
- 5 points = the overall grade of the team project (all of the students' parts added together and averaged)
- 5 points = the student's attendance (ie. involvement in the project)

Because of this grading methodology, the student's grade reflects his or her effort, or lack thereof, since 20 of the 25 points are based on the student's work. Nonetheless, 5 of the points (equaling 1/5 of the grade) depend on the other team members' efforts. This helps give every student in the group a sense of ownership in the project and encourages their involvement.

Student involvement is the other benefit of teamwork. Team projects require active learning. If students are required to work with others to write letters, memos, reports, or manuals, they cannot just sit and listen to lectures. They must put pen to paper or fingers to keyboard. We like team projects because they minimize teacher-centered education and maximize student involvement.

Teaching Suggestions

To help your students recognize the importance of technical writing, get them out of the classroom and into a work environment. Ask students to visit a job site in their field of interest. While there, have them visit several employees with different job responsibilities. These could include supervisors and subordinates from diverse work areas within a company (technicians, accountants, purchasing agents, personnel managers, etc.).

Have your students ask these people the following questions:

- What do they write (types of correspondence)?
- How much time do they spend writing on the job?
- How often do they work with teams of fellow employees on a project?
- How many employees make up the team?
- Do managers and supervisors oversee their writing?
- Will the employees provide the students samples of typical business correspondence for class discussion?

Once they return to class with samples and answers to the questions, your students can assess their findings as follows:

1. Add up the time spent writing and divide by the number of people involved. This will give the students a percentage of work time spent writing. How does the number compare to the 20 percent norm?
2. Decide which types of correspondence are written most often and in which professions.
3. Figure out the difference between the amount of time supervisors write and the amount of time subordinates spend writing.
4. Figure out how much time is spent working with others on team projects.

These answers can be decided on in small groups and then presented to the class either through brief, oral presentations or through short, written reports.

Answers to Quiz Questions

1. Technical writing differs in many ways from other types of writing; however, a primary difference is that technical writing requires a give-and-take, a dialogue, a follow-up and/or input or action.
2. The three primary considerations which help you to understand the interrelationship in technical writing include purpose, audience and tone.
3. If you're writing up to a supervisor or manager, then you're writing to request action, propose action, or recommend action. If you're writing down to a subordinate audience, then you're writing to direct action.
4. Once you recognize audience and reassess your purpose, your tone is affected. The tone for a request, proposal or recommendation written up to a manager is positive, congenial and polite. The tone for correspondence written down might be more authoritative.
5. Technical writing is a significant factor of your work experience for several reasons:
   Technical writing conducts business; technical writing takes time; technical writing costs money; technical writing is an extension of your interpersonal communication skills.
6. Good technical writing can accomplish more than just getting the job done. A well-constructed memo, letter, or report reveals to your readers not only that you know your technical field of expertise but also that you know how to communicate your knowledge thoroughly, accurately, clearly, and concisely. Through good technical writing, you reveal to your audience that you can tell people what to do and can motivate them to do it.
7. Businesses work in teams to save time and money and to ensure quality.
8. To successfully collaborate, follow this TQM strategy: Develop your team; choose a team leader to assign duties, create schedules, encourage group participation, determine team goals, identify problems, analyze the problem's causes, determine potential improvements, verify the suggested solutions, and write the text.
9. Team projects allow students to help and to learn from each other. The team projects also require that students plan; delegate responsibilities; communicate their attitudes toward word choice, tone, organization, development, and grammar; and learn to work with other people.Teamwork and collaborative writing require communication to solve problems.

CHAPTER 2

PRODUCING THE PRODUCT

- @ The Writing Process: An Overview
- @ The Rationale for Process
- @ The Process in Practice

ABSTRACT

This chapter introduces your students to the writing process. The writing process helps students achieve a high degree of success by providing them a simple, step-by-step, three-pronged, recipe approach to writing. Process allows your students to overcome the "blank page syndrome" by helping them

- ◆ ***Prewrite*** to gather information for their correspondence, organize their ideas in various way, and determine their audience's needs.
- ◆ ***Write*** rough drafts.
- ◆ ***Rewrite*** (usability testing) these drafts to improve them through revision.

Teaching Suggestions

Because this textbook integrates process into every chapter, you will want to spend time early in the semester discussing the writing process. We usually spend at least one class period explaining the fundamentals of process, emphasizing its value to students, and then reviewing a sample process log to show students what they'll be expected to do all semester.

Consider approaching this task as follows:

1. ***Provide an Overview of the Writing Process.***

The writing process is effective ... and easy! All we're asking our students to do is three things: prewrite, write, and rewrite. We tell our students that the three steps are overlapping and recursive. No one writes a rough draft without simultaneously revising, for example.

Nonetheless, to clarify what each step entails, we distinguish among the three and ask our students to construct their correspondence in three stages. Look at our sample course outlines in the "Preface" to this *Instructor's Manual.* For each assignment, students prewrite, write drafts, and help each other revise their writing through peer evaluations.

Each of these three steps should take a certain amount of time. Approximately, your students should spend 25 percent of their time prewriting, 25 percent of their time writing rough drafts, and 50 percent of their time rewriting. These percentages can help your students avoid spending ten hours writing a one page memo. We suggest that they even time themselves (or you could determine the time they spend on each stage by having them prewrite, rough draft, and revise in class). This way they could determine if they are spending too long on any one phase of the process. They also could see if they are spending the correct amount of time on revision, the most important part of the process. Students, of course, want to assume that their rough drafts are very close to a finished product. Such is not the case. Good writers revise, revise, and revise again.

Here's a personal anecdote to prove our point. You are now using our fourth edition of *Technical Writing: Process and Product.* It has been published **four times!** In doing so, it was read, reviewed, and revised by approximately three dozen editors and instructors coast to coast.

In addition, we have read and revised it numerous times ourselves. Despite these facts, we do not consider this textbook to be anything other than a rough draft—one always needing improvement. If we feel that our published book can be improved further through revision, surely your students and ours can improve a memo, letter, or report.

Do you have similar stories? Have you successfully used process to write an essay, article, editorial, poem, or letter? If so, share these stories (or even the actual documents) with your students to prove the value of process.

We emphasize the importance of process by walking around the class while students are peer evaluating.

We look over their shoulders and say,

> "I see a grammatical error in the third line of that paragraph. Shouldn't you fix it?"
> "A word in the first sentence is too long. Can't you suggest a more concise option?"
> "One of the words in your second paragraph is too vague. Isn't there a more precise way to develop your idea?"
> "Isn't that paragraph too long? Can't you find some way to make it more accessible through highlighting techniques?"

We don't give answers; we make pointed comments about the need to constantly revise.

2. *Explain Why the Writing Process is Important.*

Everyone needs a methodology for performing an activity effectively. Golf pros teach duffers how to stand, grip, and swing. Chefs teach us how to chop, dice, and saute. Homebuilders suggest ways for do-it-yourselfers to restore, refinish, and repair. Personal fitness advisors suggest ways to lose weight and build muscle.

The proven methodology for writing effectively is process. The writing process--prewriting, writing, and rewriting--offers students an easy-to-follow, easy-to-use technique for writing memos, letters, and reports.

The writing process is important because it works and because it helps students succeed. If a student is told that correspondence is due in 30 minutes, that student might be overwhelmed. In contrast, prewriting allows students to ease into a writing assignment, thereby reducing stress. Writing rough drafts allows students to "play" with different approaches to a letter or memo, rather than rushing a finished copy. This reduces stress and encourages creativity. Revising allows students to improve their first attempts at writing and to create a product they can be proud of.

3. *Provide Different Types of Prewriting Techniques.*

Prewriting helps students avoid many writing pitfalls, the worst being "the blank page syndrome." Students stare at a blank page or blank computer screen and freeze. They need to overcome this fear. Prewriting helps them accomplish this first step toward correspondence.

There's no one way to do anything. The same applies to prewriting. We want students to have a variety of ways to gather data, determine their audience, and organize their thoughts. Therefore, our text provides numerous prewriting techniques, including the following:

- reporter's questions
- mind mapping
- brainstorming/listing
- flowcharting
- outlining
- storyboarding
- cubing
- organizational charts

Each different type of Prewriting offers its unique values. Let students try different techniques so they can find the one or ones that work for them.

You probably know other types of prewriting to add to our list, such as looping, cubing, free writing, etc. Don't be limited by our choices. Use many and in different combinations.

To encourage our students to prewrite, we give them daily grade credit for submitting prewriting. In contrast, we never "grade" their prewriting. That is, we never evaluate the worth of their prewriting, assigning A's, B's, C's, etc. Doing so, we believe, would defeat the purpose of prewriting: gathering data in a non-threatening environment to help overcome the fear of writing. If our students bring prewriting to class--any kind of prewriting--we give them credit. The goal is just to get started. A finished product comes much later.

4. ***Discuss Rough Drafting.***

Before the sculptor polishes his or her statue, the cast must be made. Before the writer revises his or her correspondence, a draft must be written. Remind your students that they shouldn't worry about their correspondence while they're drafting. They don't need to spell check every word, line or paragraph during the drafting stage. Doing so slows them down and clogs their creativity. The time for sculpting, polishing, and honing the draft comes later--in rewriting.

5. ***Discuss Rewriting*** (*Usability Testing*).

In the rewriting stage, emphasize the importance of proofreading. The best organized letter on the most attractive stationery will be unsuccessful if it is laden with errors. A new problem has arisen in our classes. Students are so fascinated by graphics that they spend hours adding text art, fonts, rules, icons, and figures to their correspondence. Then they forget to spell check their writing. The report looks great, but it receives an F because of excessive errors.

To highlight this point, we have students proofread flawed correspondence. You could give your students a poorly written letter or memo, tell them that the correspondence has X number of errors, and have the students locate these errors. Use whatever samples you have handy or write a few yourself. Making errors is fun and easy.

Also, when you write correspondence with intentional errors, you can customize the errors to suit your needs. For example, what errors do your students make most often? Write new memos or letters, and include these common errors. Doing so accomplishes two goals: you help students learn to proofread, and you give them mini-grammar lessons on their commonly committed errors.

To practice usability testing, have students in peer groups evaluate a document. They could simply discuss the applicability or accuracy of the text or do hands-on testing. They then could recommend changes in the document based on their usability testing.

6. ***Review the Process Log.***

The final portion of Chapter 2 covers the writing process in practice. This chapter contains the first illustration of a process log (almost all of the subsequent chapters have process logs for the unique types of writing covered in those chapters). In this chapter, you and your students follow a problem/solution letter from inception (prewriting) through the rough draft (writing) to its revision (rewriting/usability testing). We then show the finished copy.

Ask your students to suggest additional changes to the finished copy. Remember, we believe no letter is ever perfect. Every letter can be improved.

For further discussion, we've included a complete process log in the transparency masters located at the end of this *Instructor's Manual* (see "Directive Memo from a Supervisor to Subordinates").

Conclusion

Selling the value of the writing process is easy. Just emphasize that prewriting, writing, and rewriting benefit the students. It helps them write more effectively. Prewriting gives your students a sense of direction, writing (drafting) lets them put words on the page without fear of failure, and rewriting (usability testing) gives them an opportunity to improve through revisions.

Furthermore, since these steps take time, the students have a built-in gestation period. They have time to change their minds, rethink the assignment, refine their approaches, get answers to their questions, and turn in a project they can be proud of.

Finally, the writing process helps students submit successful technical correspondence, which should earn them better grades. We can't guarantee that, of course, but it works in our classes. Because their rough drafts are read and revised by their peers, and because students have time to ask you questions, many of their errors should be caught and corrected. That gives students a chance to succeed in class. It also gives you the pleasure of reading clean and clear technical writing.

Answers to Quiz Questions

1. The three main parts of the writing process include prewriting, writing, and rewriting (usability testing).
2. In a writing activity, spend 25% of the time prewriting, 25% of the time writing, and 50% of the time rewriting.
3. Through prewriting, you can determine your objectives and your audience.
4. External motivation means that someone else has requested the correspondence; internal motivation means that you have decided to write on your own accord.
5. Before you can determine what to say and how much detail is necessary, you need to consider audience.
6. Four different prewriting techniques include answering the reporter's questions, mind mapping, brainstorming/listing, branching, flowcharting.
7. You must format your document to allow ease of access and to create a reader-friendly technical document. An unbroken page of text is not reader-friendly. To make a reader want to read, highlight key points and break up monotonous-looking text.
8. Four revision techniques include adding, deleting, simplifying, moving.
9. Failure to revise properly can cause confusion for your readers and also destroy your credibility.
10. The process approach to writing can help you to write successfully. It's generic and can be used in any writing situation, in any profession. Writing as a process will help you write any kind of technical document, for any boss, in any work situation.

CHAPTER 3

OBJECTIVES IN TECHNICAL WRITING

- @ Clarity
- @ Conciseness
- @ Accuracy
- @ Organization
- @ Ethics

ABSTRACT

What is technical writing? Your students do not know until you tell them. Technical writing differs from other types of written communication. This chapter clarifies that distinction by providing you and your students criteria for effective technical writing.

Teaching Suggestions

1. ***Clarity***—methods for developing ideas precisely.

Clarity in technical writing differs somewhat from clarity in essays or poetry. Both essays and technical correspondence should answer reporter's questions (who, what, when, where, why, and how). An expressive essay or poem, however, can clarify the writer's intent through emotional, impressionistic, connotative words. Poets and essayists can use words such as "soon," "many," "several," or "substantial." An impressionistic word, such as "near," will mean different things to different people. That's OK, in an essay or a poem, where the goal might be to convey a feeling--experienced differently by different people.

Such words, in contrast, are not acceptable in technical writing, which requires photographic, denotative words. Here's our rule of thumb: "The ultimate goal of effective technical writing is to say the same thing to every reader."

For example, let's say that I write instructional manuals for a company manufacturing space heaters. If I write, "Place the space heater *near* an open window," what will this mean to the thousands of customers who purchase the machine? One customer will place the heater 6 feet from the window, defining the word "near" accordingly. Another reader will place the heater 6 inches from the window. As the writer, I have failed to communicate clearly.

You can practice this point in your class as follows:

- Bring to class an essay or poem. Ask students to pick out impressionistic words from the readings. Compare these words to the photographic words used in technical writing to clarify the distinction.
- Ask your students individually to write a definition for the impressionistic words they found earlier. Then call on students for their definitions. Rarely will any two students have the same definition.

A simple word like "many" will be defined by one student as "4," by another as "40," and by a third as "400." Those differences might not be important to Emily Dickinson or Walt Whitman. The different opinions, however, are very significant in technical writing, where precision is the difference between safety and danger, between profit and lawsuits.

2. ***Conciseness***—methods for achieving an appropriate technical writing style.

Once again, the writing style of technical writing differs from the style of essays. Our students are familiar with essays, which encourage long paragraphs, long sentences, and long words. As teachers, we've all read student essays complete with numbers written in the margins. Students are counting words, not to limit them but to use as many as possible.

This excessive use of words is not the goal of technical writing. Good technical writing encourages short paragraphs, short sentences, and short words. One very simple example is paragraph length. As teachers, we all probably have told students "there is no such thing as a one sentence paragraph in essays." That might be true--for essays. However, many letters and memos include one sentence paragraphs. That's our point. Technical writing is different.

To prove this, give your students a sample technical report, memo, or manual. Then, provide them an essay. Ask students to compare the two types of writing based on paragraph length, sentence length, and word length. The differences should be very apparent.

3. ***Accuracy***—the importance of correct grammar and mechanics.

Correctness is important in all written communication. Grammatical or mechanical errors make writers look unprofessional and incompetent. In our opinion, this fact is even more important in technical writing than in essays.

Length of correspondence is the issue. Four or five errors in an essay are inexcusable, but they are lost somewhat because an essay can be between 500 to 1,000 words. This equals few errors in the midst of many words. In contrast, letters and memos rarely exceed 150 words. If a writer commits four or five errors in 150 words, disaster has struck.

Because we consider grammar to be so important in technical writing, here's the grading scale we use:

Number of Errors		Grade
4 major grammatical errors (in a one page assignment) We then say that if 4 major errors fails the student, 3 hurts him or her badly!	=	F
3 major grammatical errors (in a one page assignment)	=	D
2 major grammatical errors (in a one page assignment	=	C
1 major grammatical error (in a one page assignment)	=	B
0 major grammatical errors (in a one page assignment)	=	A

The student doesn't receive an A based only on grammar, of course. But the student will not receive an A with one or more errors.

When students question our grading procedures, we remind them that an A means "excellent," and that Webster's Dictionary defines "excellent" as "without flaw." No errors is how we define "without flaw."

4. ***Organization***—methods for ordering one's correspondence.

In this section, we focus attention on modes of organization, including spatial, chronology, importance, comparison/contrast, and problem/solution. Here is something that both technical writing and essays have in common.

In fact, we have included these modes of organization to help ease your transition from composition to technical writing. If you are coming to technical writing from having taught more traditional composition classes, you already know a great deal about modes of organization.

This also can help your students make the transition from composition classes to technical writing. They are used to being asked to write a comparison/contrast essay or a problem/solution essay. The same skills are required in technical writing. To emphasize this point, you can ask students to write a problem/solution memo or a comparison/contrast letter.

Here are some assignments using modes of organization.

- Write a memo or letter to your teacher, boss, parent, spouse, friend, etc. comparing/contrasting a topic. Topics could include any of the following:
  - ⇒ two movies, CDs, books--determining which to see, buy, or read
  - ⇒ two cars, trucks, bikes--determining which to buy or lease
  - ⇒ leasing vs. buying--vehicles or homes
  - ⇒ two vo-tech schools, colleges, universities--determining which to attend
  - ⇒ two employees--determining which one to hire
  - ⇒ two companies--determining which to work for

- Write a problem/solution memo or letter to your teacher, boss, parent, spouse, friend, etc., documenting a problem and suggesting possible solutions. Topics could include any of the following:
  - ⇒ problems at work--hours, pay, machinery, co-workers, parking, etc.
  - ⇒ problems at home--delegation of responsibilities, leaking roofs, peeling paint, dripping faucets, squabbling children, etc.
  - ⇒ problems with vehicles
  - ⇒ problems with spouses or children
  - ⇒ problems with friends and acquaintances
  - ⇒ financial problems

5. ***Ethics***—methods for encouraging moral standards in technical writing.

In this section, we focus on ethics from numerous angles: practical, legal, and moral. We also clarify how ethics affect technical writing by sharing with you the Society for Technical Communication's (STC) "Ethical Guidelines for Technical Communicators." Then we suggest how a technical writer achieves ethical communication by discussing STC's "Code for Communicators."

Nonetheless, ethics is a difficult topic to discuss. Ethics, by nature, are vague and colored by an individual's background. As teachers, our job is to objectify an essentially nebulous topic.

Here's what we suggest as an assignment.

1. Ask students to read our end-of-chapter case study about the Blue Valley Wastewater Treatment Plant.

2. Create a mock trial. You are the judge.

3. Select 3 or 4 students who will act as impartial jurors.

4. Divide the rest of your class into two large groups. One team will represent the defense (presenting the plant's case); the other team will represent the prosecution (presenting the North Upton community's case).

5. Select a lead counsel from each team. These people will supervise their team of lawyers.

6. Ask each team to write a brief (a short report) listing three reasons why their point of view should be accepted. The lead counsel from each team should subdivide his or her team into thirds, with each smaller group responsible for developing one of the three ideas.

7. After the reports have been written (this could take a while), ask the lead counsel to present their evidence--point/counterpoint.

8. When all evidence has been read, let the jury pass judgment on a winner.

This assignment will quantify ethics, involve students, allow them to practice critical thinking skills, and require written and oral communication. It's a fun simulation with real-world value.

Activities

Specify

1. The Accounting Department needs the fourth quarter sales reports by December 15, 2003.
2. The XYZ laser printer will replace the broken dot matrix printer in Department 23.
3. Failure to purchase a new DT 183 Oscilloscope by November 1, 2003, will delay our testing by two weeks.

4. Electrical storms in southwest Houston caused a power surge which erased accounting records from the Kordex computer system.

5. January's $500,000 decrease in sales was caused by our 20 percent loss of sales personnel due to layoffs.

6. Paper fire in the English office led to the loss of last year's sophomore research papers.

7. If we can't agree to contract terms to end the automotive technician strike by September 15, 2003, we'll lose 25 percent of our repair business.

8. The 2003 Acerbo 234 subcompact automobile has a 33 percent smaller turning radius than the 2002 model.

9. Six employees wrote letters commending her for the help she provided in writing their proposals.

10. Make sure your instructions include graphics with exploded views and callouts.

Avoid Obscure Words

1. As Very Large Scale Integration (VLSI) develops, specialized circuit simulators will be used.

2. In response to your January request, enclosed is this quarter's report.

3. Though Randolph knew he had to report any work problems, he failed to do so.

4. As you requested, the following explains our services.

5. All sales are final.

6. Because we can not find the problem, we are ending this study.

7. To start the job, here are your assignments.

8. Can you help us find what caused yesterday's machine failure?

9. Please answer the questions.

10. The car wreck was caused by faulty brakes.

Use Active Voice vs. Passive Voice

1. The Metrology Department will perform the procedure.
2. I learned two things from the conference.
3. The training supervisor explained the information and demonstrated its usefulness.
4. After discussions, the Allied representative supplied analytical equipment for automatic upgrades.
5. I also attended the Polymer symposium.
6. Vendors reviewed the process control systems for foam encapsulation.
7. To reduce overtime, let's monitor anyone working more than 48 hours.
8. Manufacturing's request led to this reassignment.
9. Our technicians should install the fiber optic network early next month.
10. Because maintenance incorrectly applied a dry lubricant, the machine broke.

Limit Paragraph Length

We use electronics to process freight and documentation. To help our customers, we are placing terminals in their export departments to track shipments. A similar system could help you monitor your exports. With a computer-generated export invoice, you could trace shipments, determine rating fees, and achieve greater efficiency. 49

Reduce Word Length

1. Tell
2. Expect
3. Find out
4. Agree, help, or assist
5. Find out
6. Try
7. Trouble or problem

8. Show or say

9. First

10. Now or soon

11. Stop, limit, or deny

12. Home, house, or address

13. Next, after, or later

14. Enough

15. Stop or end

Reduce Sentence Length

1. Monthly progress reports should be complete by the 15th.

2. Please answer these questions about your proposal.

3. Your request for a pay raise is denied. You have only worked here one month.

4. Twice, I investigated your house. Your sump pump might damage your neighbor's property. Therefore, you must fix this problem to avoid a lawsuit.

5. Let's meet to discuss replacing our old equipment.

6. The CEO's last letter stated that electrical equipment in the laboratory is most important.

7. These changes should reduce the failure rate.

8. We could implement these suggestions early next month.

9. New personnel will be tested when hired and then annually to insure they are working successfully.

10. If you have questions, please call.

Organization

For each of these activities involving organization, you could ask students to work individually or in teams. Once the writing has been accomplished, you could evaluate the results in one of two ways:

- Have the groups or individual students exchange their work, assess the results, select a winner whose work is most successful, and explain why that group's work or that individual's work succeeded.
- Pick up the writing, make transparencies of the work, review it on an overhead, and decide as a class where each piece of writing succeeded or needed improvement.

Ethics

Our suggested mock trial can involve students in a discussion of ethics.

Answers to Quiz Questions

1. Clarity is the ultimate goal of technical writing because clear writing saves time, the reader's time and the writer's time.
2. You can achieve clarity by supplying specific, quantified information, avoiding vague or abstract words.
3. Another way to achieve clarity is to answer the reporter's questions because you are addressing the who, what, when, where, why and how of the document.
4. Avoid using acronyms, abbreviations and jargon because many readers might not understand the term.
5. You can define a term either parenthetically after the term or in a glossary.
6. Concise writing can aid comprehension.
7. In a long paragraph, you force your reader to wade through many words and digest large amounts of information. This hinders comprehension.
8. A fog index is a formula to determine degree of difficulty in writing.
9. Use shorter sentences to ensure clarity and the transmission of information.
10. Wordy sentences can be caused by "shun" words, camouflaged words, and the expletive pattern.
11. Four proofreading tips include letting someone else read your document, using the gestation approach, reading backwards, and reading one line at a time very slowly.
12. If you are writing to describe the parts of a machine or a plot of ground, you might want to organize your text spatially.
13. Use chronology to document time or the steps in an instruction.
14. Ordering by importance allows you to emphasize a point.
15. Use comparison/contrast to document options and to make distinctions between two things.
16. You can use problem/solution to clarify the value of a product or a service.
17. The legal aspect focuses on liability, negligence and consumer protection laws.
18. The practical aspect focuses on the effect a document might have on the business and its success for failure.
19. The ethical aspect focuses on the morality of an issue in relation to the consumer and the business.
20. When you follow a working guide for ethical standards, you use language and visuals with precision; you prefer simple, direct expression of ideas; and you satisfy the audience's need for information, not your need for self-expression.
21. To make an ethical decision, consider defining the problem, determining the audience, maximizing values and minimizing problems, considering the big picture, writing your text.

CHAPTER 4

AUDIENCE RECOGNITION AND INVOLVEMENT

- @ **Audience Recognition**
- @ **Defining Terms for Different Audience Levels**
- @ **Multicultural Audiences**
- @ **Sexist Language**
- @ **Audience Involvement**

ABSTRACT

One of the ways in which technical writing differs most from other types of writing is its emphasis on audience. If a consumer, for example, cannot understand a user manual accompanying a purchased product, then many significant problems can arise.

1. The product could be used incorrectly, which would lead to customer dissatisfaction.

2. The product could be damaged, which would lead to customer dissatisfaction *and* force the company to repair the damage and/or provide a new piece of equipment.

3. Most importantly, the customer could be injured or killed, which would lead to lawsuits.

In contrast, if a reader fails to understand a poem, rarely does this failure to communicate lead to injury or death, nor would the poet have to replace the old poem with new, improved verse. Neither the reader nor the poet suffers significantly from a lack of understanding. The same cannot be said about technical writing, where communication is crucial.

To avoid failures to communicate, we stress that our students must recognize who their audiences are, what they know, what they don't know, and what they need to know. That demands audience recognition. Students must determine whether their audience is any or all of the following:

- high tech
- low tech
- lay
- multiple (an audience composed of the above three audience types)

We must be aware of multicultural audiences. In our global economy, students will correspond with readers for whom English is a second language. Our textbook provides guidelines to help students tackle this challenge.

We also focus on sexist language, reminding students that the work force is 50 percent female and 50 percent male. Half of their readers will be women; thus, students must be sensitive to their audience and avoid sexist language.

Different from audience recognition and sexist language, audience involvement focuses on the tone of the correspondence. We want students to achieve a personalized and positive tone in their writing, remembering that *people write to people.* To help them achieve this, we discuss

- pronoun usage
- names
- contractions
- positive words

These are used to create a more friendly, conversational, and motivational tone in letters, memos, and reports.

Teaching Suggestions

We provide techniques for communicating to different audiences. Students need to know that people from different areas of expertise (engineers, accountants, biomedical technicians, auto technicians, drafters, computer information specialists, etc.) have different levels of understanding. Despite our comments, students always want to believe that "everybody understands this abbreviation." Students assume that if they know it, others do too.

We use the following activity to prove that everyone does *not* know the same abbreviations, and that, in fact, very few people understand each other's high-tech terms:

1. Ask each student in the class to write down several acronyms or abbreviations unique to his or her area.
2. Find two or more students in class from similar disciplines.
3. Have one student ask his supposedly high-tech peer what one of the acronyms or abbreviations means. (Be sure to tell the student with the high-tech term to pick an acronym or abbreviation which will probably stump other members of the class.)

> **NOTE:** If the fellow high-tech student understands the acronym or abbreviation but others in the class are confused, then you have proven that everyone does not have the same level of understanding.

4. Ask the student to define the high-tech acronym or abbreviation.
5. Ask others in the class how many of them now know what the term means, with the aid of a definition.

> **NOTE:** Not everyone will understand the term even when it is defined. Those who understand it when it's only an acronym or abbreviation are high-tech. Those who understand it when it's defined are low-tech. Those who still don't understand it are your lay readers. They need a brief explanation.

6. Ask the high-tech students to briefly explain the term for the lay audience in class.

You can repeat this activity for other groups in class, using the acronyms or abbreviations they listed. We remind our students that everyone is simultaneously high-tech, low-tech, and lay, depending on the subject matter. If an engineer speaks to an engineer, they are high-tech peers. However, when that engineer is reading an accountant's report, the engineer now is a low-tech reader . . . or even a lay audience.

Another way we focus on the importance of audience is by having students analyze technical documents (please see exercise #1, Team Projects, in our end-of-chapter activities). You can use any number of technical documents (we find 10 documents work well). A good selection of documents for the purpose of this activity includes

- promotional brochures
- specifications
- user manuals
- technical reports
- technical articles in discipline-specific journals

After a class discussion of what constitutes writing appropriate for each type of audience, we place students in small groups to study the documents. We have found that no amount of lecture can take the place of hands-on practice to distinguish between high-tech, low-tech, and lay audiences. As they study the documents, which are passed around the room from group to group, we ask them to do the following:

1. Chart the characteristics of each document (note that we have provided a chart in our end-of-chapter activities suitable for this purpose).

2. Based on the characteristics they have found, determine which documents are intended for a high-tech audience, a low-tech audience, or lay readers.
3. Group by group, discuss which documents they have designated high, low or lay, adding their reasons for doing so.

Students are very good at distinguishing among different types of technical documents. Students also are very opinionated. This activity generates a great deal of involvement, discussion, debate, and critical thinking.

Activities

1. Avoiding Sexist Language

a. All software development specialists and their spouses (or significant others) attended the conference.

b. The supervisors met to discuss techniques for handling union grievances.

c. Technicians must keep accurate records for their monthly activity reports.

d. The president of the corporation met with her sales staff.

e. Throughout history, scientists have always tried to make their mark on society through discoveries of significant intellectual worth.

f. The chairperson of the meeting handled the debate well.

g. All fabricated (or hand-made) components were checked for microscopic cracks.

h. The supervisors have always bought flowers for women employees on Mother's Day.

i. Mr. Smith (the CEO) introduced his staff: Mr. Jones (the vice president), Mr. Brown (the chief engineer), and Mrs. White (the executive secretary).

j. The police officers and the firefighters rushed to the scene of the accident.

2. Audience Involvement

a. Before we process your request, John, we would like more information.

b. To improve mechanical safety, please implement this new procedure.

c. Thank you for your efforts. Your department supervisor appreciates your help.

d. Sue, you've done a great job. You've surpassed this month's quota by 12 percent and deserve our thanks.

e. When your computer overloads, simultaneously press "control" and "reset." If the screen then reads "data recovered," you can continue your operations. However, if the screen reads "1/0 error," please call your computer resource center.

f. Congratulations Fairway Village! Ampex Corporation announces the opening of a new office in your neighborhood.

g. Telech's new communication system can help you meet your business telecommunication needs. Our new system offers multi-party extensions, call forwarding, call waiting, intercom, and call accounting.

h. You could achieve a higher rate of successful manufactured products through line checks, system checks, and/or random checks.

i. If you would listen to my speech, you could help me improve my content and delivery.

j. To number your pages, please follow these steps: If you want to number from page one, press "page layout" and then the number "1." To start numbering from any other page, press "page layout" and then type in the page number you want your document to start on.

3. Reader Benefit

a. Please sign the attached waiver allowing us to lay your cable.

b. John, by purchasing the correct program, we will be able to keep efficient records and maintain positive customer relations.

c. To maintain your services, please pay the enclosed bill.

d. To achieve high grades, quote and paraphrase correctly.

e. Please send the information by January 12. Doing so will allow us to service your account more efficiently.

f Correct input of information ensures timely shipments, reduces unneeded expenses, and leads to increased year-end raises.

g. Thank you for your recent call about our product.

h. Your team has won 6 of their last 12 games.

i. Thank you for your suggestions. The reconfiguration you suggest is too large for the area specifications. If you would like to submit new calculations, we would be happy to reconsider your ideas.

Team Projects

Please see our earlier comments ("Teaching Suggestions").

Answers to Quiz Questions

1. The four main types of audience include high-tech, low-tech, lay, and multiple.
2. High-tech readers work in your field of expertise and share your educational background, work experience, or level of understanding. Low-tech readers include your co-workers in other departments or your bosses, your subordinates, or your colleagues who work for other companies. Though they might share your type of education, they might be unfamiliar with your company's procedures or in-house jargon, acronyms, and abbreviations. Customers and clients who neither work for your company nor have any knowledge about your field of expertise are your lay audience. Multiple audiences include high, low and lay readers.
3. Three definitions include words or phrases that define your term parenthetically or in a glossary, sentences that define your term either in a glossary or following the unfamiliar term, or an extended definition of one or more paragraph.
4. The main parts of a definition include a term, the type, and distinguishing characteristics.
5. In today's global economy, your company will market its products or services worldwide.
6. Follow these guidelines to write correctly to a multicultural audience: define acronyms and abbreviations, avoid jargon and idioms, distinguish between nouns and verbs, watch for cultural biases/expectations, avoid slash marks, avoid humor or puns, leave space for a translation, avoid figurative language.
7. Sexist language expresses a bias based on gender.
8. Avoid sexist language by using appropriate pronouns, by treating women and men equally, by avoiding stereotyping, by choosing nonsexist nouns.
9. You can achieve personal tone through pronoun usage, names and contractions.
10. You humanize the text when you use pronouns.
11. You create a friendly reading environment in which you speak directly to your reader when you use names.

12. You can express reader benefit and ensure audience involvement by (1) explaining the benefit to the reader and by (2) using positive words and verbs.

→ a car is a vehicle that is driven on land.

a ___ is a ________ that (verb) ________.

HyperText Transfer Protocol

computer access code

(provides)

secure communications on the Internet, an intranet, or an extranet.

— be careful w/ #s measures, dates, time

Use stylized

UNIT 2
CORRESPONDENCE

CHAPTER 5

MEMOS

- @ The Differences among Memos, Letters, and E-mail
- @ Criteria for Writing Successful Memos
- @ Process
- @ Sample Memos

ABSTRACT

Every corporate employee writes memos daily. Memos (and their electronic counterpart, e-mail--discussed in Ch. 13) are written up to supervisors, down to subordinates, and laterally to co-workers for a host of reasons: to ask questions, direct duties, report on activities accomplished, propose new approaches to job duties, compare options, argue positions, and so on. Because of their frequency and versatility, writing effective memos is an essential skill for students.

This chapter introduces your students to prewriting, writing, and rewriting techniques for memos. We provide you objectives, criteria, samples, process logs, and end-of-chapter activities to help you teach memos.

Teaching Suggestions

We always use collaborative writing projects in class (please see our discussion on collaborative writing in Ch. 6 of this instructor's manual). These group projects give our students an opportunity to practice writing and to receive immediate feedback on their successes and failures.

You could have your students write group memos about any topic of your choice. The students could be asked to write a group informative memo, a problem-solution memo, a comparison-contrast memo, an argument-persuasion memo, etc. The possibilities are unlimited.

To set the stage for such memos, we create fictitious scenarios--mini case studies. For example, you could tell your students that they are a social committee at work designated to plan a company picnic. First, they must prewrite using mind mapping to gather data about this picnic. Then they must write the rough draft of an informative memo providing the company agenda for this picnic. Finally, the group must rewrite the draft, using revision techniques discussed in the chapter.

As another example, teach your students criteria for a memo. Then, tell them that they have just attended a training session on memo writing offered by a world-renowned training facilitator. Now they must return to their companies and write a memo to their subordinates, telling the workers how to write a memo. As in the preceding scenario, the students must prewrite to gather data, write a group rough draft, and revise the draft through rewriting.

These collaborative writing activities allow students to practice each of the three steps in the writing process. You can create similar scenarios more attuned to your students' unique needs. Our end-of-chapter activities provide additional ideas.

You might want to show other unedited memos to your students:

1. Show the unedited sample to your students.
2. Ask your students for comments regarding the memo's successes.
3. Ask your students for revision suggestions to improve the memo's problems.
4. Show your students the edited version and compare their suggestions with yours.

Activities

This chapter's activities require audience assessment. Assignment #1 gives you numerous options for memo topics, and each topic implies a different audience. Some topics require the writer to play the role of a boss writing to a subordinate; some require the writer to play the role of a subordinate writing to a boss. Be sure your students recognize the importance of audience as it relates to tone and the required level of technicality.

Team Projects

Following is a sample response to assignment #3.

Date: April 6, 2003
To: Jan Pascal
From: Wilkes Berry
Subject: REQUEST FOR ENGINEERING SERVICES

From January to March, Lenexa City Hall has received approximately 36 phone calls from residents complaining about odors from city sewers. We have been hired to look into this problem.

Please ask your engineers to perform the following tests on Lenexa's sewer system:

- corrosion
- water seepage
- pipe decay
- alignment
- depth
- soil composite

We need these studies completed by May 1, 2003. If you have any questions or would like further information, please call me (ext. 3625). A successful job on this project could mean more work in Lenexa. Let's all perform professionally.

Answers to Quiz Questions

1. Memos are written to someone within a company.
2. The five distinctions among memos, letters and email include destination, format, audience, topic and tone.
3. A subject line is important because 100% of your audience reads this line; this is where you begin talking to your reader.
4. A subject line consists of a focus, a preposition, and a topic.
5. A goal of a good introduction is to tell your readers what you want and why you're writing.
6. Be concerned with document design in your memo to ensure reader-friendly ease of access.
7. Three highlighting techniques in a memo include itemization, white space, and headings.
8. Conclude your memo with a complimentary close and/or a directive close.
9. Your audience is high or low-tech in-house readers.
10. Your writing style in a memo depends on conciseness, clarity and accessibility. Use simple words, readable sentences, specific detail, and highlighting techniques.
11. Clustering/mind mapping allows you to see graphically the relationship between subpoints and components of an idea.

CHAPTER 6

LETTERS

- @ Letter Components
- @ Letter Wizards and Templates
- @ Letter Formats
- @ Criteria for Different Types of Letters
- @ Process

ABSTRACT

Chapter 6 introduces your students to prewriting, writing, and rewriting techniques geared toward seven different types of letters:

- ◆ Inquiry
- ◆ Cover (Transmittal)
- ◆ Good News
- ◆ Bad News
- ◆ Complaint
- ◆ Adjustment
- ◆ Sales

The goal of this chapter is to familiarize your students with the different types of letters they might encounter on the job, and to provide them with techniques that will help them communicate successfully. The prewriting method we offer in this chapter (Reporter's Questions), along with the drafting approaches and the skills required for rewriting are especially appropriate for letters. Other chapters in the textbook offer different prewriting, writing, and rewriting concerns geared toward different types of correspondence.

Teaching Suggestions

Collaborative Writing

Students learn how to write by writing, just as people learn how to putt by putting, to paint by painting, or to shoot free throws by shooting free throws. We all need practice to perfect skills. Therefore, we turn our classrooms into writing workshops as often as possible. Here's the process we follow in teaching our classes:

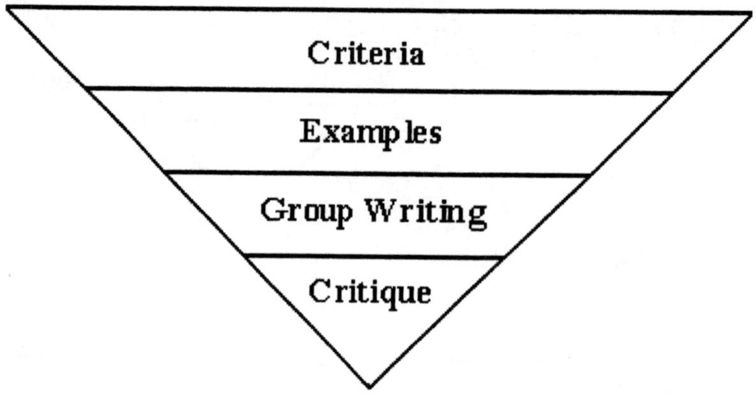

After spending time explaining objectives and criteria for the letter we have assigned, we discuss the writing process and review examples of both student-written and professionally-written letters. Then we allow our students to practice what we have been preaching through group writing.

> **I**n our experience, the best way to practice the writing process is through collaborative writing activities--best for the students, best for the instructors. If each student writes an individual letter using prewriting, writing, and rewriting, you will have difficulty helping him or her. None of us has enough classroom time to assist every student.
>
> **C**ollaborative writing, however, minimizes the number of letters you read and maximizes your teaching effectiveness.

We divide our students into groups of three (two students aren't a group; four students are unwieldy). We give all the groups the same writing assignment. For letters of inquiry, for example, we set up this scenario:

- You have just graduated from college (congratulations) with a BA.
- You want to pursue an MA degree to make yourself more marketable.
- Write a letter of inquiry to a university of your choice.
- In this letter, ask five questions about the MBA program or university.

We don't tell our students which questions to ask. That's the essence of their assignment. In the group letter, they must determine what information they need.

To accomplish this end, the students must prewrite (answering the reporter's questions), write a rough draft, and revise the letter through rewriting. We then critique their group work, providing them immediate feedback. You could provide your students feedback in two ways:

- Gather the group letters and make transparencies for class review.
- Have the groups exchange their group letters and provide each other feedback.

> Students benefit from collaborative writing activities. They help each other, they learn from each other, they acquire hands-on practice employing your criteria for letter writing, they use the writing process to generate their text, and they receive immediate feedback on their writing. This approach demands active learning versus passive listening.

You could try similar collaborative writing activities for any of the other types of letters discussed in this chapter. Ask your students to

- Create a new product or service and write a sales letter advertising their creation.
- Respond negatively to a client's claim in a bad news letter.
- Congratulate a subordinate for his or her achievement in a good news letter.
- Write a cover letter prefacing a multiple page document that you bring to class.
- Write a letter of complaint to a company regarding their flawed product.
- Respond to that letter of complaint in an adjustment letter.

In each instance, have your students prewrite, write, and rewrite. The entire activity (criteria, examples, group writing, and critique) usually takes approximately 1 to 2 hours.

Activities/Case Studies

You can use our end-of-chapter activities and case studies either for in-class collaborative work or for the students' out-of-class assignments. Since each of the letter writing assignments at the end of this chapter requires a different audience, you might want to use this as an opportunity to discuss audience needs (see Ch. 4). Ask your students to determine which letters require high-tech to high-tech writing, high-tech to low-tech writing, etc. Decide which letters could use contractions effectively and which letters would avoid such casual writing.

These letter assignments also provide your students an opportunity to role play. Have students write a sales letter followed by a letter of inquiry, or a bad news letter followed by a good news letter. In one instance, they would be a high-tech inventor selling a product. Then they could write the letter of inquiry as a low-tech client needing more information. Perhaps the bad news letter could come from a customer complaining about a purchased product. A high-tech manager could write a follow-up good news letter offering the customer satisfaction.

To further assist your students in their prewriting, you might want to use the following **REPORTER'S QUESTIONS CHECKLISTS** for

- sales letters
- cover letters
- good news letters
- bad news letters
- adjustment letters
- complaint letters

(A reporter's questions checklist for letters of inquiry is in the textbook.)

COVER LETTER--REPORTER'S QUESTIONS CHECKLIST

1. **Who** is your audience?

    - High-Tech Audience ____________________
    - Low-Tech Audience ____________________
    - Lay Readers ____________________

2. **What** is the topic of your letter (the general subject matter of your attachment)?

    ____

3. **Why** are you writing this letter (why are you sending the enclosed information)?

    ____

4. **What exactly** is within the enclosure of special interest to your reader(s)? Itemize these components and tell your reader(s) how to find the information--page numbers or section numbers.

    ____

    ____

    ____

    ____

    ____

5. **What's** next?

    - When do you plan a follow-up? ____________________
    - Why is that date important? ____________________

GOOD NEWS LETTER--REPORTER'S QUESTIONS CHECKLIST

1. **Who** is your audience?

   - High-Tech Audience ____________________
   - Low-Tech Audience ____________________
   - Lay Readers ____________________

2. **What** is the topic of your letter (the good news)?

   ____

3. **Why** are you writing this letter?

   ____

4. **What exactly** are the causes or effects of this good news? Itemize these points for your reader(s).

   ____

   ____

   ____

   ____

   ____

5. **What's next**?

   - When do you plan a follow-up? ____________________
   - Why is that date important? ____________________

BAD NEWS LETTER--REPORTER'S QUESTIONS CHECKLIST

1. **Who** is your audience?

   - High-Tech Audience ____________________
   - Low-Tech Audience ____________________
   - Lay Readers ____________________

2. **What** buffer can you provide your reader(s) as a lead into the bad news?

   ____

3. **What exactly** are the facts that support and lead up to this bad news? Itemize these points for your reader(s).

   ____

   ____

   ____

   ____

   ____

   ____

4. **What's the bad news**

   ____

5. **What's next?**

   - What alternatives or options can you suggest for your reader(s)?

     ____

   - How can you end your letter upbeat?

     ____

SALES LETTER--REPORTER'S QUESTIONS CHECKLIST

1. **Who** is your audience?

    - High-Tech Audience ____________________
    - Low-Tech Audience ____________________
    - Lay Readers ____________________

2. **What** is the topic of your letter (the product or service you are selling)?

    ____

3. **Why** will this product or service help your reader(s)?

    ____

4. **What exactly** are the benefits of this product or service? Itemize these points for your reader(s).

    ____

    ____

    ____

    ____

    ____

5. **What's next?**

    - How will you make your reader(s) act?

      ____

LETTER OF COMPLAINT--REPORTER'S QUESTIONS CHECKLIST

1. **Who** is your audience?

    - High-Tech Audience ________________
    - Low-Tech Audience ________________
    - Lay Readers ________________

2. **What** is the topic of your letter (the subject matter of your complaint)?

    ____

3. **Why** are you writing this letter (what has gone wrong)?

    ____

4. **What exactly** are your reasons for this complaint, or what exactly do you want from your reader(s)? Itemize these points.

    ____

    ____

    ____

    ____

    ____

5. **What's next?**

    - When do you want an answer? ________________
    - Why is that date important? ________________

LETTER RESPONDING TO COMPLAINT--REPORTER'S QUESTIONS CHECKLIST

1. **Who** is your audience?

    - High-Tech Audience ____________________
    - Low-Tech Audience ____________________
    - Lay Readers ____________________

2. **Why** are you writing (100 percent accepted, 100 percent rejected, or partially accepted)?

    ____

3. **What** buffer will you provide for a rejection?

    ____

4. **What exactly** will you tell your reader(s) to justify the response? Itemize these points for your reader(s).

    ____

    ____

    ____

    ____

    ____

5. **What's next?**

    - How will you end your letter upbeat?

      ____

Answers to Quiz Questions

1. Letters are external correspondence that you send to a colleague working at another company, to a vendor, to a prospective client, to an agency, or to a friend who lives around the corner or across the continent. Letters leave your work site.
2. Your letters not only reflect your communication abilities but also are a reflection of your company. If your letter is sloppy, is marred with grammatical errors, or creates a negative tone, you will look bad and so will your company. Your company can be held legally responsible for your written mistakes.
3. The eight essential letter component's are the writer's address, the date, the inside address, the salutation, the letter body, the complimentary close, the signed name and the typed name.
4. Four optional letter components include the subject line, new page notations, writer's and typist's initials, and an enclosure notation.
5. If you are seeking information or answers, you will write a letter of inquiry.
6. Until you tell your readers who you are and what you want to know, they can't know.
7. A cover letter lets you tell readers up front what they are receiving and it helps you focus your readers' attention on key points within the enclosures.
8. You might write a letter promoting an employee or offering an individual a job at your corporation; these are examples of good news letters.
9. Conciseness and clarity in a bad-news letter's introduction might be harsh and abrupt; you might anger your reader and keep them from reading the letter. To avoid these lapses in diplomacy, begin your bad-news letter with a buffer.
10. You might include supporting details in a letter of complaint introduction such as serial numbers, dates of purchase, invoice numbers, check numbers, names of salespeople involved in the purchase.
11. Three ways to adjust a letter of complaint include 100% yes, 100% no, and a partial adjustment letter.
12. You might try a variety of lead-ins to a sales letter such as an anecdote, a question, a quotation, or data.
13. In a sales letter, you should use strong verbs, format for reader-friendly ease of access, audience recognition with pronouns, ease of use, and implied urgency.

CHAPTER 7

THE JOB SEARCH

- @ How to Find Jobs
- @ Criteria for Resumes
- @ Sample Resumes
- @ Criteria forLetters of Application
- @ Sample Letters of Application
- @ Techniques for Interviewing
- @ Sample Follow-up Letters
- @ Checklists

ABSTRACT

If our students don't find employment, we are all in trouble. Long unemployment lines will result, and society will suffer. In this chapter, we provide numerous techniques for helping your students find jobs.

Teaching Suggestions

Finding a job is paramount. This is an important chapter to cover in any technical writing course. In fact, many technical writing journals suggest that teaching resume writing skills and letters of application might be the ideal way to begin a technical writing class. After all, resumes and letters of application certainly require clarity, conciseness, accessibility, audience recognition, and accuracy. What better way to emphasize the importance of these key technical writing concerns than in correspondence that might one day help your students make money?

A great teaching unit would consist of the entire job package:

- a job search (via newspapers, networking, college placement centers, or online searches)
- a letter of application
- a resume
- a mock interview (using oral presentation skills discussed in Chapter 18)
- a follow-up thank you letter

You could facilitate this unit in two ways: theoretical or actual. For example, you could teach the job package theoretically just as a mock, in-class activity. Peer groups could represent simulated hiring committees. The process could look as follows:

1. Individual students could write resumes and letters of application.
2. These student applicants could submit their job package to the "hiring committee" for review.
3. The student committee could review the resume and letter of application and respond to the student applicants with their feedback. This way, the student writers could receive objective, diverse viewpoints regarding their job packages.
4. The committee could invite the applicants in for an interview, providing the student applicants a chance to polish presentation skills.
5. The committee could "hire" one individual, explaining their rationale for this choice. In doing so, they would clarify for the students not hired what went wrong and how to improve for subsequent, actual job searches.

You also could teach this unit as an actual job search. To do so, follow this procedure:

1. Either through newspaper ads, networking, your school's placement center, or online searches, students should find a real job for which they might want to apply.
2. Students write resumes and letters of application, appropriate for the job to which they are applying.
3. Students either mail their job package to the company for review, or--for better control within a classroom environment--submit the resume and

application to one of your school's job placement employees or to a personnel employee from a company you've contacted.

4. Students wait for a response from the company, or--for better control within the classroom environment--your personnel employee could review the resume and letter of application and respond to the student applicants with feedback. This way, the student writers could receive objective, professional assessments regarding their job packages.

Activities

Using our end-of-chapter activities will give your students a better understanding of the job search.

Answers to Quiz Questions

1. Four places you can look for a job include visiting your college or university job placement center, talking to your instructors, networking with friends or past employers, and checking with professional affiliations and publications.
2. The primary goal of your resume is to get you an interview.
3. Two types of resumes include the reverse chronological resume and the functional resume.
4. Write a reverse chronological resume if you are a traditional job applicant, hope to enter the profession in which you have received college training or certification, have made steady progress in one profession, and plan to stay in your present profession.
5. Your career objective informs the reader of your resume's focus.
6. List jobs in reverse chronological order.
7. In reverse chronological order, document your education including previous schools, colleges, military courses and/or training seminars.
8. Optional inclusions in the education section include grade point average, academic club memberships and leadership offices held, unique coursework, special class projects, academic honors, percentage of college expenses you've paid, or technical equipment you can operate.
9. If you are changing professions or reentering the workforce after a long absence, you will want to write a functional resume.
10. Exclude personal data such as birthdate, race, sex, religion, height, weight, religion, marital status, or pictures of yourself.
11. Avoid sentences in the resume because too many will begin with "I"; you might commit grammatical errors; and they'll take too much room.
12. For ease of access and to highlight, bulletize in a resume.

13. Be specific and use examples to "prove you're great."
14. To design a resume effectively, you can capitalize or boldface headings, indent subheadings to create white space, underline subheadings and italicize major accomplishments, use bullets to create lists, change font sizes, etc.
15. Your online resume will allow thousands of international employers to access it instantly; allow you to include portfolios of your work with lengthy documents, full-color graphics, animation, and sound; allow you to update it quickly and easily; and allow you to show your expertise with HTML and Web-ready graphics.
16. A resume is generic whereas a letter of application is specific, customized for that job.
17. Use a call to action in the conclusion to the letter of application, and mention your resume.
18. To interview successfully, dress for success; be on time; don't chew gum, smoke, or drink coffee; watch your body language; and control your speaking.
19. Your follow-up letter allows you to thank your interviewer, keeps your name fresh in their memories, and gives you an opportunity to introduce new reasons for hiring you.

UNIT 3

VISUAL APPEAL

CHAPTER 8

DOCUMENT DESIGN

- @ Chunking
- @ Order
- @ Access
- @ Variety

ABSTRACT

People choose to read short stories, poetry, plays, novels, newspapers, and magazines. When they read these types of communication, they do so at their leisure and at their pace. None of the above scenarios apply to technical writing. Employees don't choose to read memos, letters, and reports. They are required to do so as part of their jobs. Employees don't read technical documents at their leisure or at their pace. They must read e-mail, memos, letters, and reports rapidly and in the midst of doing their other work.

Due to these facts, employees need to be given information in a manner that allows them to skip, skim, and scan for details--rapidly. That's where document design comes in. It's not enough just to dump massive amounts of data on your readers. Instead, technical writing must be visually designed to allow readers to find information easily. In this chapter, we offer numerous techniques for reformatting a page.

Teaching Suggestions

A technical writer's success is dependent on how well the document is designed. To clarify this point, we ask our students to imagine this scenario:

"It's 5:00 p.m. You're about to go home after a long day's work. Just as you are about to leave, your boss comes into your office and hands you an envelope. Because you're a conscientious employee, you open the envelope and here's what you see."

Then we show them Figure 8.1, a poorly designed page. Instantaneously, our students groan. That's the correct response. A poorly designed page is an immediate turn off. It's almost a visceral experience.

With this image in their minds, we have explained to them why page layout is important. Next, we can show them how to improve the document's design, using the various techniques discussed throughout the chapter. Figures 8.2 through 8.7 reveal these options.

Students will want to use <u>all</u> their new-found highlighting techniques on every page of every document. That's a mistake. Writers should not clutter a page with too many highlighting techniques. We tell our students that the highlighting techniques used to achieve document design are like spices for the stew. A few spices add flavor; dumping the entire pantry of spices onto a page makes the stew indigestible. A little goes a long way.

Activities

3. *Reformatted Memo*

Option I

DATE:	November 30, 2003
TO:	Jan Hunt
FROM:	Tom Langford
SUBJECT:	CLEANING PROCEDURE FOR MANUFACTURING WALK-IN OVENS #98731, #98732, and #98733

The above mentioned ovens need extensive cleaning. To do so, please follow this procedure:

1. Vacuum and wipe all doors, walls, roofs, and floors.

2. Remove all vents/dampers.

3. Tack cloth the vents/dampers to remove loose dust and dirt.

4. Replace all filters.

Doing the above twice a week will keep wet parts free of dust and dirt and reduce extensive rework.

Option 2

DATE:	November 30, 2003
TO:	Jan Hunt
FROM:	Tom Langford
SUBJECT:	CLEANING PROCEDURE FOR MANUFACTURING WALK-IN OVENS #98731, #98732, and #98733

The above-mentioned ovens need extensive cleaning. To do so, please follow this procedure:

ACTION	ITEM						
	Doors	Walls	Roofs	Floors	Vents	Dampers	Filters
Vacuum	X	X	X	X			
Wipe	X	X	X	X			
Remove					X	X	
Tack					X	X	
Cloth							
Replace							X

Doing the above twice a week will keep wet parts free of dust and reduce extensive rework.

4. *Reformatted Executive Summary*

EXECUTIVE SUMMARY

Objectives

The City of Waluska wants to provide its community a safe and reliable water treatment facility. We have two primary goals:

- Protecting Waluska's environmental resources
- Ensuring Waluska's community values

Achieving These Goals

Hardtack and Sons (H & S) Engineering can achieve these goals. We understand the project and recognize your community's needs. Specifically, we can update Waluska's Loon Lake Water Treatment Plant (LLWP), providing Waluska techniques which will allow the City to

- Manage changing regulations and protection standards
- Develop financially responsible treatment services
- Plan land use for community expansion
- Uphold the community's values

The H & S Difference

Our company understands water treatment technology and knows your community.

Company Credentials

For the last 10 years, H & S has taken a leading role in waste water treatment. We have created feasibility studies for Loon Lake toxic control, developed odor abatement procedures for your streams and creeks, and assessed your water treatment plant's ability to meet regulatory standards. Julie Schopper, our Partner in Charge, has worked on similar projects worldwide and demonstrated her leadership through effective communication skills.

Community Knowledge

H & S personnel are members of your community. Your concerns are their concerns. Our dynamic project team has a close working relationship with your community's regulatory agencies.

Conclusion

H & S offers the City of Waluska an integrated program addressing all your community's needs. We believe H & S is your best choice to ensure that your community receives a waste water treatment plant ready to meet the challenges of the 21st Century.

Answers to Quiz Questions

1. Carefully design a technical document so it is reader-friendly, has accessible content, appeals to the reader, and represents your company positively.
2. Breaking text into smaller chunks of information is chunking.
3. Chunking helps your readers to understand the overall organization of your correspondence.
4. You can achieve chunking with headings, white space, rules, and section dividers and tabs.
5. You can help your audience prioritize information.
6. You can achieve a hierarchy or headings with type size, density by boldfacing, spacing, and positioning.
7. Hierarchical headings allow the reader to visualize the order of information, to see clearly how the writer has prioritized text.
8. Help your reader to access information with white space, bullets, numbering, boldface, all caps, underlining, italics, windowing, fills, color, and inverse type.
9. You can use too many highlighting techniques.
10. A few highlighting techniques help your readers filter out background data and focus on key points; too many are a distraction and clutter the document design.
11. Four ways to vary document design include printing horizontally, using more columns, varying gutter width, and using ragged-right margins.

CHAPTER 9

GRAPHICS

- **Color**
- **Criteria for Effective Graphics**
- **Three-dimensional Graphics**
- **Types of Graphics**

ABSTRACT

Readers cannot live by word alone. People need tables and figures to enliven the text and clarify information. In this chapter, we explain why and when graphics are used--to aid conciseness, ensure clarity, and add cosmetic appeal. We provide criteria for the different types of graphics, including tables and figures. We give numerous samples for class discussion, and we offer end-of-chapter activities, allowing students to practice their graphics skills.

Teaching Suggestions

We rarely teach graphics as a stand-alone assignment. Graphics (tables and figures) are used not to replace text, but to enhance text in technical descriptions, instructions, short reports, long reports, and oral presentations. To accomplish this goal, you could follow this procedure:

1. Teach a type of correspondence (description, instruction, report, presentation). Then, explain how a graphic could aid communication. Doing so provides the students a context.

2. Show students where graphics have been used in our chapters on descriptions, instructions, reports, and presentations. You could discuss placement of graphics, along with the text that leads into and follows the graphics.

3. Provide criteria for the different types of graphics. This not only includes the difference between tables and figures, but also the different types of figures—pie charts, line graphs, bar charts, photographs, schematics, flow charts, line drawings, Gantt charts, icons, geologic maps, topographical charts, CAD drawings, combination charts, etc.

This is the most important step in discussing graphics. There are many different types of graphics, but all are not effective in every situation. Now's a good time to distinguish among the different types and explain when a pie chart works better than a bar chart or line graph. More importantly, when is a simple bar chart more effective than a 3-D bar chart? Computer programs provide students with an overwhelming array of graphic options. Every program comes complete with a gallery of graphics (six different types of bar charts, six different types of pie charts, etc.). Though computers allow us to create scatter charts, Venn diagrams, and histograms, it's not always in our best interest to do so. Students need to select the graphic which communicates most clearly, not the one that is most gaudy.

4. Use our end-of-chapter activities to practice creating different types of graphics.

5. Bring to class your own examples of instructional manuals, year-end reports, sales brochures, etc. Discuss with your students which graphics are used, how they are placed in the text, and whether they are effective or not. Have students explain why the effective graphics succeed and why the poor ones fail. Then ask students to replace the flawed graphics with improved ones.

Activities

1. Presenting information in a bar chart, pie chart, and table

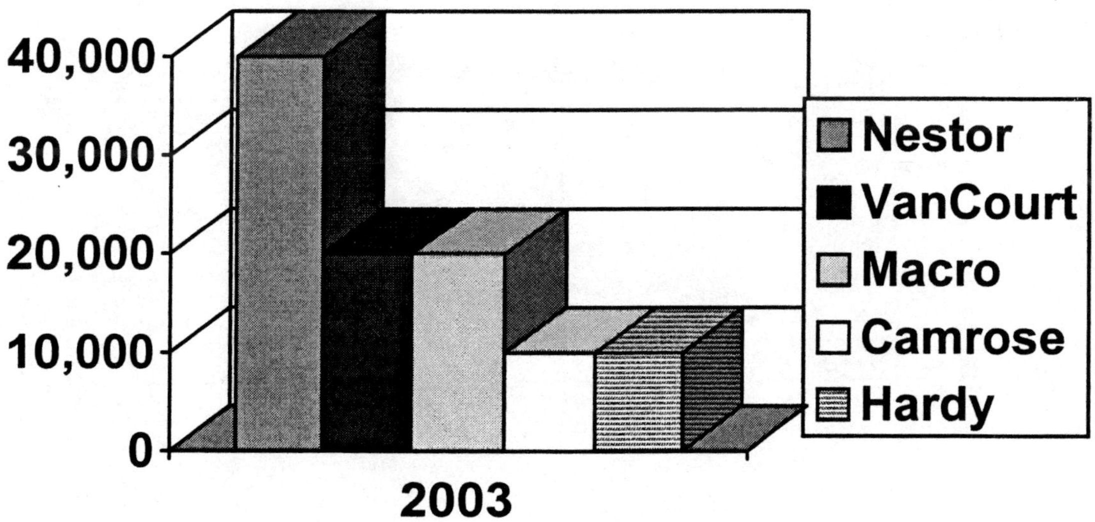

Figure 1 2003 Relay Purchases

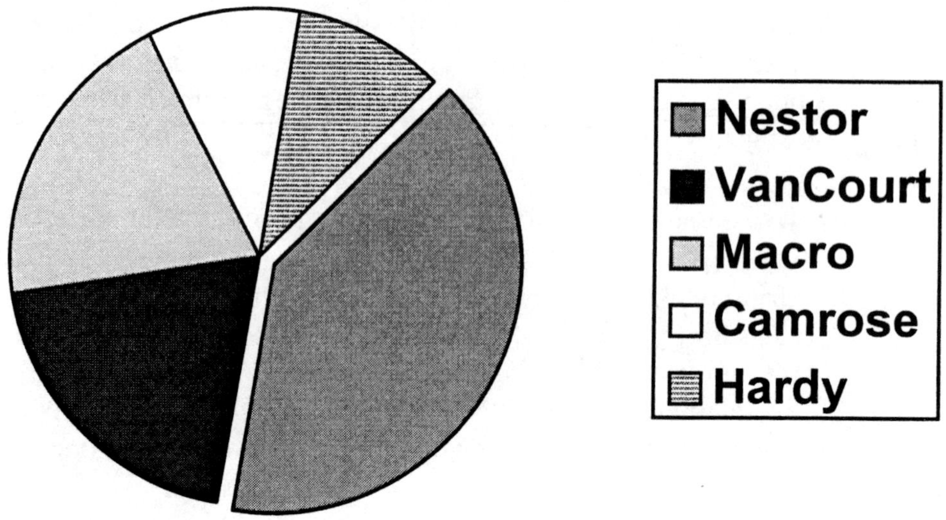

Figure 2 2003 Relay Purchases

Table 1 2003 Relay Purchases	
RELAY TYPES	NUMBERS PURCHASED
Nestor	40,000
VanCourt	20,000
Macro	20,000
Camrose	10,000
Hardy	10,000
Total	**100,000**

2. Presenting information in a grouped bar chart and a table.

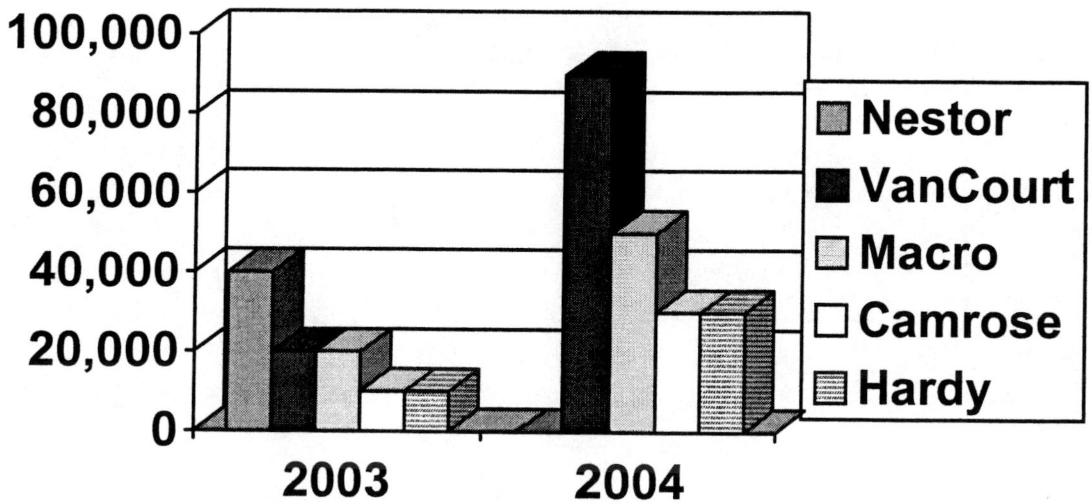

Figure 1 2003 vs. 2004 Relay Purchases

Table 1 2003 vs. 2004 Relay Purchases		
RELAY TYPES	**NUMBERS PURCHASED IN 1999**	**NUMBERS PURCHASED IN 2000**
Nestor	40,000	0
VanCourt	20,000	90,000
Macro	20,000	50,000
Camrose	10,000	30,000
Hardy	10,000	30,000
TOTAL	**100,000**	**200,000**

3. *Create a table*

Table 1 Frequency Voltage Calculations

FREQUENCY(Hz)	VOLTAGE (V)
3K	8
2.8 K	9
2.3 K	10
1.8 K	11
1.4 K	12
1.14 K	12.5
926 Hz	13

Case Study

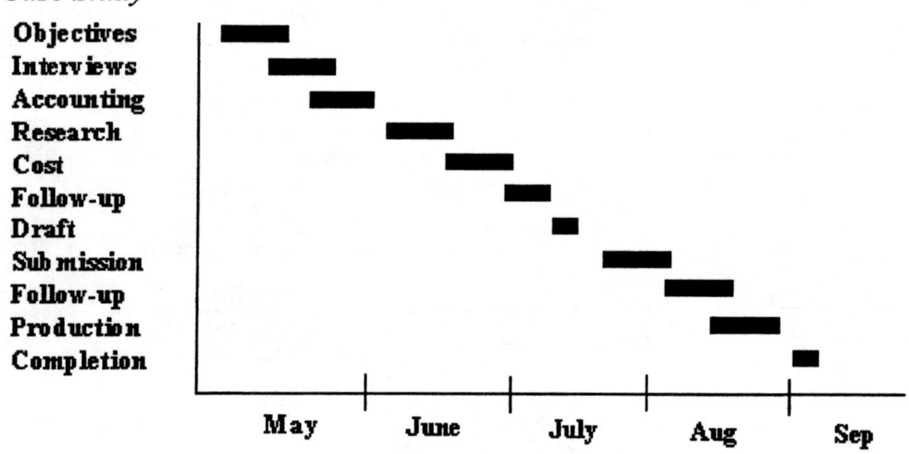

Figure 1 Project Schedule

Answers to Quiz Questions

1. To present large blocks of data and/or reveal comparisons, you can supplement, if not replace your text with graphics.
2. Visual aids allow you to provide large amounts of information in a small space.
3. Visual aids can clarify and show trends, comparisons between like components, percentages, and facts and figures.
4. Visual aids help you break up the monotony of wall-to-wall words.
5. Avoid using color if it is distracting, if the color is too "high-style," if the color increases production costs, if they take up too much memory, and if you're concerned that the color might be responded to differently by different readers.
6. A 3-D graphic does not convey information quantifiably.
7. Successful visuals are integrated into text, are appropriately placed, add to the material in the text, communicate important information easily, and are not distracting.
8. Tables show comparisons and highlight the content's significant differences.
9. After a table, explain its significance.
10. Be consistent when using numbers. Use either decimals or numerators and denominators for fractions.
11. Figures highlight and supplement important points in your writing.
12. Five types of figures include bar charts, pie charts, line charts, flowcharts, and organizational charts.
13. Provide a legend or key at the bottom of a figure to explain information.
14. Two types of bar charts are horizontal bars and vertical bars.
15. A pictograph uses picture symbols instead of bars to show quantities.
16. To create effective pictographs, the picture should be representative of the topic discussed, each symbol equals a unit of measurement, and more symbols of the same size indicate a higher quantity.
17. Gantt charts show chronological activity.
18. Pie charts illustrate portions of a whole.
19. Too many wedges in a pie chart crowd the chart and confuse readers.
20. A flowchart shows chronological sequence of activities.
21. Schematics are useful for presenting abstract information in technical fields such as electronics and engineering.
22. Icons are visual representations of a capability, a danger, a direction, an acceptable behavior, or an unacceptable behavior.
23. To create graphics for your web site, download existing online graphics, modify and customize existing online graphics, or create new graphics.

UNIT 4
TECHNICAL APPLICATIONS

CHAPTER 10

FLIERS, BROCHURES, AND NEWSLETTERS

- @ Why Write Fliers?
- @ Criteria for Writing Fliers
- @ Why Write Brochures?
- @ Criteria for Writing Brochures
- @ Effective Brochure Usability Checklist
- @ Why Write Newsletters?
- @ Criteria for Writing Newsletters
- @ Process
- @ Effective Newsletter Usability Checklist

ABSTRACT

This chapter introduces your students to the exciting world of fliers, brochures, and newsletters. Creating these documents allows writers to bring together all their technical writing knowledge of clarity, conciseness, document design, accessibility, tone, and highlighting techniques.

Why write a flier, brochure, or newsletter? Corporations and companies; educational institutions; churches and synagogues; museums, zoos, and amusement parks; cities and municipalities—they all need to communicate with their constituencies (clients, citizens, members, etc.). Sometimes a letter or e-mail doesn't provide enough space for ample communication. A report might be too stiff and unfriendly. That's where fliers, brochures, and newsletters come in handy.

With a one or two-page flier, the writer can briefly convey information about an upcoming event, new product, or improved service. If more development is needed, the writer can use a brochure. To update readers weekly or monthly or quarterly, maybe a newsletter will better serve the purpose. These types of communication are valuable options for writers when other types of communication just don't quite suit the purposes.

This chapter introduces your students to objectives, criteria, samples, process, usability, and end-of-chapter activities to help you teach fliers, brochures, and newsletters.

Teaching Suggestions

Because of time limitations in a college semester, you might only have enough time to choose one of these documents as a class project. We find that any of these three documents works well when written in small groups (writing of these types of documents in business is often done in small groups so it's useful for students to have this experience in your technical writing class).

One way to begin this small group writing activity is to give students a case study that provides detailed information about a company (we provide you a case study in this chapter).

In addition, you can allow your students to create their own topics. Consider these options for each the types of correspondence.

@ Fliers

Fliers will be short, focusing on only one key aspect of a topic. This could include a new and unique feature of a service or product, for example. Ask your students to brainstorm topics that would fit such an assignment. These could include new features for

- software or hardware
- an e-mail system
- a VCR or stereo
- their vehicle
- a PDA (palm pilot)
- a home appliance
- a service

The list is endless. Once students find a limited topic, they can proceed to the following:

1. Create a title: Ask a question to arouse reader interest and focus on the reader's need.

2. Determine the text: Your students won't need much, perhaps a sentence or two, a few bulleted points. This gives them a chance to practice conciseness and limit their focus. The text should highlight the product or service's most unique features.

3. Select a graphic: This will not only entice the reader, but also be informative.

4. Find the phrase: What catchy phrase will the readers remember and associate with the product, service, and/or company?

5. Design the flier: They'll focus on layout, color, font, etc.

6. Make it correct: Grammar!

@ Brochures

Brochures will be longer than fliers and shorter than newsletters. Again, the topics are endless.

Students can write brochures about places, people, things, or services. These include brochures about

- a high school's National Honor Society, drama club, and athletic teams
- a city's entertainment options
- a park's lecture series on wildlife
- a college's degree program for Computer Information Services
- vacation spots; a museum's special exhibit; an amusement park's shows and show times; historic homes and people; flora and fauna in a regional area; tire makes and models; etc.

Your students will only be limited by their imaginations.

Students can get a good idea of what brochures look like by bringing samples to class.

1. Ask each student to find 5-6 sample brochures about various topics.
2. Once the students have brought these to class, ask them to assess each brochure.

Which ones are most successful, and why? Which ones fail, and why? Based on these decisions, the students should have a better idea of what their brochures should look like.

Again, we like team projects. An especially good project for brochures, one close at hand, is the city in which you live. Think of the possibilities--restaurants, zoos, museums, banks, arboretums, schools, retirement homes, businesses, and parks! Send the students out on field trips to visit these sites. They can take pictures, interview representatives, gather data through research, and then write their brochures. Have them follow the steps we've provided for the process. Then they can assess their successes with our usability checklists.

@ Newsletters

Newsletters provide the most detailed information of the three types of correspondence discussed in this chapter. And again, the topics are endless. Think of the newsletters you get or have gotten: from your children's schools (K-12 and college); from your alma mater; from your church or synagogue; from your investment brokerage firm; from your place of employment; from one of your professional affiliations.

Bring these and others to class for your students to review, or have them bring examples of their own to class. Review the samples to determine which are successful, and why, which are unsuccessful, and why.

Students can then brainstorm their own topics, use our prewriting, writing, and rewriting techniques to create the newsletter, and then assess it according to our usability checklist.

Whether you choose to have your students write fliers, brochures, or newsletters, consider following this approach:

1. Give the class the criteria to follow for the document you want them to write--either a flier, a brochure, or a newsletter.
2. Show them the samples we have provided in the textbook.
3. Ask students to share other samples they have found on their own.
4. Divide the class into small groups (3 to 4 students work well together).
5. Applying your criteria for the assignment, ask the students to rough draft their document.
6. Give students time to work in a pc lab because all of these documents need the enhancement of design and graphics.
7. Ask students to share their rough drafts with another group.
8. Make sure students revise and rewrite their original draft based on the feedback they received from their peer-evaluating group.
9. Ask students to share their document with the entire class.

This collaborative writing activity allows students to practice their writing and receive immediate feedback. In addition, the group writing in the classroom simulates the real world of business group writing that students will encounter in business. Because students will work in teams on the job, we can give them practice in the classroom to practice teamwork.

@ Interaction with other teams

You can also have your students decide which of the fliers, brochures, and newsletters they've created does the best job of communicating to its intended audience. They can choose the best of the documents and explain why this is the best example. They should base their decision on the criteria you provided them to begin their assignment.

Answers to Quiz Questions

1. In a flier you can include information about a company's products or services.
2. To access electronic fliers, brochures, or newsletters, use an Internet browser such as Yahoo that lists these online sources by topic.

3. Four benefits to using a flier are that it is cost effective, time efficient, responsive to immediate needs of an audience, and personalized with a specific market or client in mind.
4. You should emphasize only one key point per flier.
5. Font size and type can be used effectively in a flier to make text more readable and more dramatic.
6. To highlight your flier, use bullets, white space, tables, boldface, italics, and headings.
7. Use a catchy phrase to personify your company's primary focus and to capture your company's personality.
8. If you have more information to convey than can fit on a one-page flier, a brochure might be a good option.
9. Write a brochure to create awareness of your company, product, or service; to increase understanding of a product, service, or your company's mission; to advertise new aspects of your company, product, or service; to change negative attitudes; or to show ways in which your company, product, or service surpasses your competition.
10. On a title page of a brochure, include the topic, a graphic, and contact information.
11. On the back panel of a brochure, you can include a conclusion, mailing information, coupons, and/or locations.
12. Headings and subheadings in a brochure act as navigational tools to guide your readers, direct their attention, and help them find the information they need.
13. In a brochure, use document design effectively by limiting sentence length, using white space, using color, bulletizing key points, boldfacing, and using graphics.
14. When considering audience in a brochure, avoid problems by defining any terminology, abbreviations, and/or acronyms that could cause confusion.
15. Cubing is a prewriting technique that allows you to look at your topic from six angles or perspectives: describe your topic; compare/contrast your topic; define your topic; associate your topic; analyze your topic; and argue the pros and cons of your topic.
16. Cubing is so effective because it allows you to apply a variety of senses to a topic and to analyze it from multiple perspectives. Thus, you can gather a great deal of information.
17. In a brochure, you could discuss benefits, prices, locations, uses, specifications, options, warranties, etc.
18. Usability determines whether something meets a user's needs.
19. Usability focuses on four key factors: retrievability, readability, accuracy, and user satisfaction.
20. You can write about many things in a newsletter such as information about a school or a business; rules and regulations of a homeowner's association; membership benefits, etc.
21. A newsletter consists of a flag, columns, headlines, subheadings, pop-up quotes, sidebars, and a table of contents.

22. In a newsletter, use highlighting techniques such as tables and figures, font changes, bullets, numbering, boldface, and italics.
23. Drawing a storyboard sketch of your newsletter will help you plan your layout, length, and graphical needs.

CHAPTER 11

TECHNICAL DESCRIPTIONS

- @ Criteria for Writing Technical Descriptions
- @ Process
- @ Process Log

ABSTRACT

You work as a purchasing agent for a company. Your job is to buy office equipment, including chairs, desks, file cabinets, book cases, and so on. To do so, you review a sales catalogue provided by a potential vendor. While reading this catalogue, you see that the chairs come in various colors, sizes, shapes, and materials. What you're reading is a technical description of that company's products.

You have a position on the city council. An engineering company has submitted a proposal suggesting how they will repair your city's bridges. As part of their proposal, they explain the current problems with these structures and show how their new bridges will be superior. They list the materials they will use, provide the various tolerances and load bearing capabilities, and visually depict the bridges' construction. What you're reading is a technical description of that company's products.

Technical descriptions (or specifications), a major part of many businesses, are used to sell, explain, document, and report. This chapter provides criteria for technical descriptions, samples for your students to emulate, a writing process your students can use to create their descriptions, a process log which models the writing process, and end-of-chapter activities for practice.

Teaching Suggestions

Technical descriptions lend themselves well to collaborative writing activities (please see our discussion of collaborative writing in Chapter 6 of this instructor's manual). By having your students write in groups, they practice what you've been lecturing, they help each other write, and they learn from each other.

The best type of in-class collaborative writing activity for technical descriptions involves simple tools, mechanisms, or pieces of equipment--things that can be brought to class, handed around the room for review, and described in a limited amount of time. These could include shoes, glasses, coffee cups, pens, spiral notebooks, staplers, a computer mouse, used floppy disks, and so on. The key is to find everyday items which the students might have on them and which have identifiable parts. As we remind our students, each of these items was manufactured by some company and then sold. Prior to this manufacture and sale, some person or team of people wrote technical descriptions for the items. It was their job. The manufacturing personnel then followed the technical specifications to mass produce the objects.

Our favorite item is a pencil. It has four, easy-to-identify parts: eraser, metal band, wooden body, and lead. Here's the sequence we follow when having our students describe a pencil:

1. Divide our class into small groups.
2. Hand each group a pencil.
3. Allow the groups to gather detail through prewriting (listing)--color, material, shape, size, etc.
4. Ask the groups to write a rough draft (including graphics).
5. Ask the groups to revise their drafts.
6. Pick up the group-written descriptions and review them. We either make transparencies for class review or have groups exchange with each other.

This collaborative writing activity allows students to practice their writing and receive immediate feedback. In addition to this value, the group writing creates project teams, simulating the importance of TQM (Total Quality Management) and CQI (Continuous Quality Improvement) in the work place. Our students will be asked to work in teams on the job. We can help them succeed later by providing them an opportunity now to practice teamwork.

In a technical description activity, each team has its graphic artist (the person designated to sketch the pencil), a technical specialist (the person in charge of measurements), and a technical writer (the individual who writes the rough draft). All members of the group are responsible for quality control--rewriting for accuracy.

Graphics

Speaking of graphics, technical descriptions offer students an ideal opportunity to practice integrating text and pictures. For our group-written pencil description, we ask students to sketch a pencil (either free hand or using a computer graphics program). In this sketch, they must include labeled callouts and either an exploded view, cutaway view, or a cross-sectional view of the pencil. To help them understand these concepts, we refer them to Ch. 9.

Students are nervous about the graphics we require. When they see computer-generated graphics, they can't imagine how they will be able to create graphics of this caliber. To ease their stress, we remind them that this is a technical writing class, not an art class, and that we don't expect professional-quality graphics. However, we do expect their graphics to be conscientiously created. Artistic ability is one thing; neatness is another.

We give them three options for creating graphics: hand-drawn graphics, computer-generated graphics, or clip art (graphics "borrowed" from another source and pasted on their text). If the students do use graphics from elsewhere, we require that they document their sources to avoid plagiarism. This, by the way, can lead to a discussion of ethics.

Answers to Quiz Questions

1. A technical description is a part-by-part depiction of the components of a mechanism, tool, or piece of equipment.
2. A technical description provides the end user of a manual with information about the mechanism's features or capabilities.
3. A technical description is comprised of a title, a discussion section, graphics, call-outs, and a conclusion.
4. The super comic book look is large, easy-to-follow graphics that complement text.
5. Call-out labels identify key components of a mechanism.
6. Three places for graphics include text and graphics separate, text and graphics merged, and text and graphics merged with an exploded view.
7. Use spatial organization
8. Include for development weight, size, materials, identifying numbers, color, shape, density, make/model, texture and capacity.
9. Photographic words are denotative, quantifiable, and specific. Impressionistic words are vague and connotative.

CHAPTER 12

INSTRUCTIONS AND USER'S MANUALS

- @ Criteria for Writing Instructions
- @ Process
- @ Process Log
- @ Criteria for Writing a User's Manual

ABSTRACT

We published an article which documented the results of a nationwide survey sent to teachers of technical writing and practicing technical writers ("A Survey of Technical Writing Practitioners and Professors: Are We on the Same Page." *Proceedings: 42nd Annual Society for Technical Communication Conference, pages 44-47).*

In this survey sent to over 700 professionals coast to coast, we asked 72 questions regarding types of writing and writing skills. One of the categories we focused on was the most often mentioned types of writing taught and/or practiced on the job. **The most often mentioned type of writing for both teachers and writers was instructions.**

Ninety percent of writers polled listed instructions first; 87 percent of teachers polled listed instructions first. As is clearly evident from this survey result, instructions are an essential type of writing for teachers to teach and students to practice.

Teaching Suggestions

As with all our topics for technical writing, we have our students practice writing in collaborative groups (please see our discussion of collaborative writing in Chapter 6 of this instructor's manual).

Since almost every classroom has an overhead projector in it, we use this readily available machine as our topic for a collaboratively written instruction. Any topic, however, would work. You might want to bring tinker toys, erector sets, Lincoln logs, refillable fountain pens, programmable calculators, etc. to class for use as topics for instruction. If your class has computers, they would provide an excellent topic: how to print, save, copy, or change font; how to create graphics; how to set tabs or margins; etc.

For our how-to-operate instruction for the overhead projector, we follow this procedure:

1. Form small groups (three to four students).
2. Set the overhead projector in the center of the room for observation and testing.
3. Allow the groups to gather detail through prewriting (listing and flowcharting)—what steps one takes to operate the overhead.
4. Ask the groups to write a rough draft (including graphics).
5. Ask the groups to revise their drafts.
6. Pick up the group-written instructions and review them. We either make transparencies for class review or have groups exchange with each other.

Our textbook provides you a sample of this project from prewriting through the finished product.

Audience Recognition

No type of technical writing requires audience recognition more than instructions. A successful instruction speaks directly to the needs of its audience. An unsuccessful instruction provides either too much data or too little information, failing to recognize its audience's needs. We remind our students that an unsuccessful instruction, one failing to provide its audience enough information, can lead to severe problems: the equipment can become damaged, and/or individuals can be hurt. Both these problems can be costly for the company. On one hand, equipment must be replaced, often at the company's expense. On the other hand, lawyers' fees, medical costs, and personal damages must be paid. Both situations can hurt a company significantly. The writer has a legal, practical, and ethical responsibility to focus on these issues.

NOTE: This is an excellent time to discuss ethics from a very practical standpoint.

One way to emphasize the importance of audience in instructions is to review examples. You might want to bring to class instructional manuals for electronics, computers, automotives, woodworking, etc. Ask your students to determine what audience was intended (high-tech, low-tech, lay) and whether the information provided was on target. If the information was correct, ask your students to explain how the writers succeeded. That is, what techniques did they use successfully? If the information was not sufficient, ask your students to suggest what more was needed. The added detail could be better graphics, more parenthetical definitions, or more detailed explanations for each step.

Activities

7. *Revising a flawed instruction.*

DATE: November 30, 2003
TO: Maintenance Technicians
FROM: Second Shift Supervisor
SUBJECT: CLEANING PROCEDURE FOR MANUFACTURING WALK-IN OVENS *#98731, #98732,* and *#98733*

The above mentioned ovens need extensive cleaning. To do so, please follow this procedure:

1. Vacuum and wipe all doors, walls, roofs, and floors.
2. Remove all vents/dampers.
3. Use a tack cloth on the vents/dampers to remove loose dust and dirt.
4. Replace all filters.

Doing the above twice a week will keep wet parts free of dust and dirt and reduce extensive rework.

Answers to Quiz Questions

1. Use instructions to show how to operate a mechanism, install equipment, manufacture a product, package a product, or unpack equipment.
2. Three parts of well-organized instruction include an introduction, discussion, and conclusion.
3. In the introduction, include the topic, your reasons for writing, and the number of steps or sections involved.
4. Use hazard alerts to avoid potentially harming the user or to avoid costly lawsuits.
5. Place hazard alerts before the step.
6. Use highlighting techniques to make your hazard alert accessible.
7. *Caution* warns of the potential for damage or destruction of equipment. *Warning* says there is the potential for serious personal injury. *Danger* says there is the potential for death.
8. Use a note for important information.
9. *Caution* is yellow, *warning* is orange, and *danger* is red.
10. In a hazard alert, include the hazard alert word, color, alert symbol, identification of the hazard, consequence, avoidance steps, and a hazard icon.
11. Organize instructions chronologically.
12. A disclaimer says a company is not responsible.
13. When you consider audience, you are more careful to clarify and not to assume.
14. Graphics clarify your points.
15. Use short sentences to avoid overloading your reader.
16. Begin each step with a verb.
17. Approximate usability testing by letting a team of peer evaluators evaluate your manual.
18. Features of usability testing include audience recognition, development, conciseness, consistency, ease of use, and layout.

UNIT 5
ELECTRONIC COMMUNICATION

CHAPTER 13

E-MAIL, ONLINE HELP, AND WEB SITES

- @ The Internet—The "Information Superhighway"
- @ The Intranet—A Company's Internal Web
- @ The Extranet—A Web Within a Web
- @ E-mail
- @ Online Help
- @ Web Sites

ABSTRACT

We are in the midst of a communication revolution. Times are changing (or they already have changed). Communication for the masses began with drum beats and fire signals, progressed to hieroglyphics on sandstone and elaborate calligraphy in medieval bibles, and then was conveyed to many via the Gutenberg press.

The next stages of mass communication involved the telegraph, radio, telephones and then television. We have now entered a new realm of communication--cyberspace. Whatever we feel about the relative merits of the Internet, it's here to stay, and it will shape irrevocably the future of how our students communicate. If we want to help prepare our students to succeed on the job, we must focus on electronic communication, which will be an integral part of their careers.

This chapter gives you an overview of three different types of electronic communication: e-mail, online help, and Web sites. We provide you criteria for communicating in cyberspace and numerous end-of-chapter activities which allow your students to practice these important skills.

Teaching Suggestions

Our textbook provides information about traditional, paper-bound correspondence, including memos, letters, resumes, reports, and proposals (Chapters 5, 6, 7, 16, and 17). We also focus attention on effective technical writing style, including conciseness, clarity, audience, and document design (Chapters 3, 4, and 8). All of this information is needed to construct successful electronic communication. Effective e-mail, online help, and Web sites require conciseness, clarity, page layout, and audience recognition. Each of the traditional, paper-bound types of correspondence (instructions, resumes, reports, technical descriptions—as well as forms and directories) can be transformed into online documents.

To teach electronic communication, ask your students to change their paper text to hypertext or their memos to e-mail. By doing so, you can show that all effective writing requires similar techniques and skills.

For example, have your students do the following:

- *Reformat a resume or a short report as Web pages.* To do so, suggest that the students use their first-level headings in their paper-bound text as home pages, and their second- and/or third-level headings as hypertext links (hot buttons). These links could either create pop-ups or jumps for additional information. This would minimize the size of the screens, keeping readers from having to scroll through numerous pages.

- *Reformat an instruction for Web pages.* To do so, suggest that your students use their first-level headings, in their paper-bound instruction as home pages, and their steps for the instructions as hypertext links. These links, along with any cautions, warnings, or notes, could lead either to pop-ups or jumps for additional information. An added challenge for your students would be graphics. Adding graphics to online documentation requires students either to copy and save the graphics from existing online documents (this, of course, raises an ethical issue), or to create their own graphics. We discuss online graphics in Ch. 9.

- *Access corporate home pages on the Internet.* Once the students access these home pages, they could analyze them, focusing on document design, clarity, conciseness, audience recognition, and audience involvement. Students could discuss why the successful Web pages succeeded and why the unsuccessful ones failed. Topics of discussion could include the use of color, highlighting techniques, hypertext links, sales tone, etc. Then the students could use the successful Web pages as models for their own online documents.

- *Reformat a paper-bound memo to e-mail.* Doing so would allow your students to practice the economy required for cyberspace correspondence. In addition, students could learn the difficulties created with such highlighting techniques as boldface, italics, or iconic bullets, which are not always achievable for e-mail.

- *Communicate with other students in class or at remote sites via e-mail.* Doing so would allow students to practice the economy required for cyberspace correspondence and to focus on tone.

- *Define terms using online help screens.* Choose high-tech terms from students' areas of interest and/or majors. Then, write extended definitions of these terms as online help screens. Students could then compare their definitions with similar ones found online either in computer software programs or on the Internet.

Activities

Hypertext

4. Revision of Figure 9.6

Home Page

PROBLEMS CAUSED BY POWER SURGES AND BROWNOUTS

Your company, located in the East Side Commercial Park, is subject to power surges and brownouts. These are caused when industrial motors and environmental equipment frequently start and stop. Your copiers also cause power surges.

Due to these surges and outages, your company suffers three types of costly problems:

1. **Damage to electronic equipment.**
2. **Lost data and failed backups and file transfers.**
3. **Staff time losses.**

Hypertext Link #1

PROBLEMS CAUSED BY POWER SURGES
AND BROWNOUTS

Your company, located in the East Side Commercial Park, is subject to power surges and brownouts. These are caused when industrial motors and environmental equipment frequently start and stop. Your copiers also cause power surges.

Due to these surges and outages, your company suffers three types of costly problems:

1. **<u>Damage to electronic equipment.</u>**

Power surges and brownouts have damaged your three network servers, power supplies, hard drives, memory, and monitors. The repair costs <u>for these</u> damages totaled $55,000.

Hypertext Link #2

PROBLEMS CAUSED BY POWER SURGES AND BROWNOUTS

Your company, located in the East Side Commercial Park, is subject to power surges and brownouts. These are caused when industrial motors and environmental equipment frequently start and stop. Your copiers also cause power surges.

Due to these surges and outages, your company suffers three types of costly problems:

1. **Damage to electronic equipment.**

2. **Lost data and failed backups and file transfers.**

Not being able to shut down the servers during power failures is leading to system crashes. Data loss requires time-consuming tape restoration. If the failures happen before a scheduled backup, the data is lost. This requires new data entry time. Ten percent of your data loss can not be restored. This will lead to loss of customer confidence.

Hypertext Link #3

PROBLEMS CAUSED BY POWER SURGES AND BROWNOUTS

Your company, located in the East Side Commercial Park, is subject to power surges and brownouts. These are caused when industrial motors and environmental equipment frequently start and stop. Your copiers also cause power surges.

Due to these surges and outages, your company suffers three types of costly problems:

1. **Damage to electronic equipment.**
2. **Lost data and failed backups and file transfers.**
3. **Staff time losses.**

Due to power surges and brownouts, you spent $40,000 on time losses caused by network downtime. Data entry staff can not work while servers are being repaired. The staff then must spend extra time reentering lost data. Fifteen percent of the Data Management staff time is now spent recovering from power-related problems.

Answers to Quiz Questions

1. The internet is a global information system.
2. An intranet is contained, including several linked local area networks.
3. An extranet is a collaborative network that uses internet protocols to link businesses with their suppliers, customers, or other businesses that have common goals.
4. Email is quick, cheap and efficient instant communication.
5. Problems with email include poor or incomplete documentation, computer limitations, lack of privacy, and misunderstandings and/or erroneous messages.
6. To create effective email, use the correct email address, limit your email message to one screen, organize your email, provide an effective subject line, and use a few highlighting techniques.
7. Netiquette is appropriate online behavior.
8. To achieve pleasant email, be courteous, don't write abusive or angry messages, learn how to express emotions visually, and read and reread your message before you send it.
9. Online help is increasing because of the increased use of computers, the reduced dependence on hard-copy manuals by consumers, the need for readily available online assistance, and proof that people learn more effectively from online tutorials.
10. Pop-ups and jumps give more information about your topic such as overviews, processes, definitions, procedures, examples, cross references, and tutorials.
11. To achieve effective online help screens, organize your information for easy navigation, recognize your audience, achieve a positive and personal tone, design your document, be concise, be clear, and use correct grammar.
12. To achieve a pleasant tone, encourage users, be constructive, not critical, write in the affirmative, and use personal pronouns.
13. To design a help screen, use color sparingly, be consistent, use a 10-point sans serif font, and use white space.
14. Reading a screen takes longer than reading printed text.
15. Successful web sites have good margins, a logical sequence, and little visual "noise."
16. Too much sound or visual distractions can cause "noise" and make your web site hard to read.
17. The home page is the first page on your web site.
18. A good home page includes identification information, a graphic, a lead-in introduction, and hypertext links/buttons.

19. Hypertext links/buttons allow the reader to click and jump to another page.
20. The linked pages should contain headings/subheadings and development.
21. To achieve effective design on a web page, use a nice background, graphics, highlighting techniques, a pleasant style and tone, and correct grammar and punctuation.
22. Limit use of italics and underlining.
23. HTML coding is a computer language used for coding.
24. You don't need to know HTML coding to design a web page because you can rely on software.
25. Knowing HTML coding gives you control over the editing of your web page.

UNIT 6
REPORT STRATEGIES

CHAPTER 14

RESEARCH

- @ Criteria for Writing Research Reports
- @ Process

ABSTRACT

In this chapter, we help students locate information in the library or online, read and analyze sources of information for use in technical correspondence, write research reports, and document sources of information (using either MLA or APA formats).

Teaching Suggestions

We use this chapter on research much as we use our chapter on graphics, as an enhancement to other types of technical writing. We rarely teach research as a stand-alone assignment. Instead, we teach research as follows:

- To help students develop ideas for use in short reports, long reports, summaries, and/or oral presentations.
- To help students with their career searches. Researching a job or a field would help students understand career opportunities, pay scales, and job requirements.
- To help students understand the value and uses of on-line research through an Internet browser, such as Yahoo or Gopher. Using the Internet for research would give your students a better understanding of Web page construction, on-line resumes, corporate sales literature, instructions, and/or the use of hypertext links within a document or to additional research sources.

In addition to the above suggestions, you will find the following rules helpful when teaching your students how to quote and/or paraphrase.

Rules for Quoting

Following is a paragraph as it appears in our Chapter 12: Instructions and User's Manuals.

> Even pop-top cans of soda include instructions: "I. Lift up. 2. Pull back. 3. Push down." When such simple items as soup and soda need instructions, you can imagine how necessary instructions are for complex products and procedures.

To quote any or all of this paragraph, you should follow these rules:

1. Place all quoted material within quotation marks.

   "Even pop-top cans of soda include instructions: '1. Lift up. 2. Pull back. 3. Push down.' When such simple items as soup and soda need instructions, you can imagine how necessary instructions are for complex products and procedures."

2. Use double and then single quotation marks if your material to be quoted already contains quotation marks.

   "Even pop-top cans of soda include instructions: 'I. Lift up. 2. Pull back. 3. Push down.' When such simple items as soup and soda need instructions, you can imagine how necessary instructions are for complex products and procedures."

3. Provide a parenthetical source citation following the quote.

   "Even pop-top cans of soda include instructions: '1. Lift up. 2. Pull back. 3. Push down.' When such simple items as soup and soda need instructions, you can imagine how necessary instructions are for complex products and procedures" (Gerson and Gerson 240).

4. Make sure that every word within the quotation marks is *exactly* as it appears in the original source. You can not change a word, add a word, or omit a word inadvertently.

   Incorrect
   "Even a pop-top can of soda include instructions: 'I. Lift up. 2. Pull back. 3. Push down.' When such simple items as soda need instructions, you can imagine how necessary instructions are for complex products and procedures" (Gerson and Gerson 240).

   In this example, we used quotation marks correctly, but we quoted incorrectly. We changed words and omitted words.

5. Limit the number of words quoted. Twenty-five words per quote is a preferred maximum. Quotations should be used to support a writer's ideas, not to replace the student's writing. An excessive number of words quoted is not effective.

6. To limit the number of words quoted, quote only key words and phrases.

7. Since key words and phrases can not stand alone, incorporate these key words and phrases in a sentence of your own construction.

   Example
   You must include instructions when explaining "complex products and procedures" (Gerson and Gerson 240).

8. Supply a lead-in to your quotation.

   Example
   Gerson and Gerson say that "complex products and procedures" require written instructions (240).

9. Use ellipses(...)--three spaced periods--to show omission of words within a quote.

   Example
   "Even pop-top cans of soda include instructions: 'I. Lift up. 2. Pull back. 3. Push down.' When simple items need instructions, you can imagine . . . how necessary instructions are for complex products and procedures" (Gerson and Gerson 240).

10. Don't ignore basic rules of grammar when quoting. Just because you've used quotes does not mean that rules of grammar no longer apply.

   Incorrect
   After saying that "Even pop-top cans of soda include instructions," Gerson and Gerson state the following, "When such simple items as soup and soda need instructions, yo can imagine how necessary instructions are for complex products and procedures" (240).

   This is a comma splice, a type of run-on sentence. To correct it, place a colon after the word "following."

Rules for Paraphrasing

Distinguishing between quoting and paraphrasing is also helpful for your students. You can remind them that they can avoid using only quotations by paraphrasing. A fine line exists between correct paraphrasing and plagiarism. When a person plagiarizes (intentionally or unintentionally), he or she either fails to rewrite the author's sentences completely or fails to provide a parenthetical source citation after the paraphrase.

To avoid these problems, teach the students to paraphrase correctly as follows:

1. Do not add information within a paraphrase which is not stated explicitly in the author's original words.
2. Rewrite the original information completely, using your own words and sentence structure.
3. Don't place quotation marks around the paraphrase.
4. Provide a parenthetical source citation.
5. Make sure the paraphrase is approximately the same length as the original source of information.

The following paragraph is taken from our Chapter 3: Objectives in Technical Writing.

> The ultimate goal of good technical writing is clarity. If you write a memo, letter, or report which is unclear to your readers, which your readers can't understand, then what have you accomplished? You've wasted time. First, your readers don't understand your point or can't follow your train of thought. They must write you a follow-up inquiry to determine your needs. This wastes *their* time. Once you receive the inquiry, you must rewrite your correspondence, trying to clarify your initial intentions. You've now written twice to accomplish the same goal. This wastes *your* time.

Your correct paraphrase might look like this:

> Your primary goal in technical writing should be clarity. Whatever type of correspondence you generate, writing clearly will save time. For example, if you write a memo or letter clearly, your readers won't have to waste their

> time writing you back, asking for more information. Similarly, you won't have to waste your time responding to their inquiry (Gerson and Gerson 25).

On the other hand, here is an example of a flawed paraphrase:

> The primary goal of good technical writing is clarity. If you write a memo or letter which is unclear to your readers, what have you accomplished? You've wasted their time and yours. First, your readers don't understand your point, they must write you a follow-up inquiry to determine your needs. This wastes *their* time. Once you receive the inquiry, you must rewrite your correspondence to clarify your intentions. This wastes *your* time (Gerson and Gerson 25).

This writer has correctly supplied a parenthetical source citation, but the supposed paraphrase is plagiarism for two reasons. First, the sentence structure is almost exactly the same as the original. Second, many of the words and phrases are the same and should have been quoted. In essence, the writer has stolen words, phrases, and syntax from the original source.

Rules for Summarizing

We discuss summaries thoroughly in Chapter 15, but here's a brief overview of what summarizing entails:

1. Rewrite the original information in your own words (just like paraphrasing).
2. Make sure that the summary is a condensed version of the original. For example, one sentence could summarize a paragraph.
3. Cite the source of your information parenthetically after the summary.

Following is a summary of the same paragraph used to discuss paraphrasing:

> A clearly written memo or letter saves time both for the reader and the writer since neither party will have to rewrite for further information or explanation (Gerson and Gerson 25).

Activities

Using our end-of-chapter activities will give your students a better understanding of research.

Answers to Quiz Questions

1. You can research information in online catalogues, reference books, indexes for technical subjects, CD-ROMS, and internet search engines.
2. Considering audience allows you to determine the amount of technical definition necessary for effective communication.
3. A solicited research report is informative; an unsolicited research report is persuasive.
4. Research reports are more formal than other types of technical writing.
5. Use effective formatting and highlighting techniques to create a reader-friendly researched report.
6. A focus statement guides you and controls your research.
7. On search engines, you can research any topic you can imagine.
8. Document your sources to avoid plagiarism and to give credit to the writer.

CHAPTER 15

THE SUMMARY

- @ Criteria for Writing Summaries
- @ Process
- @ Process Log

ABSTRACT

A primary goal of effective technical writing is conciseness. People in a work environment don't have the time or the desire to read excessively long correspondence. One way to achieve this desired conciseness is to summarize. This chapter will help your students understand how to, when to and why to summarize.

Teaching Suggestions

Our technical writing course is packed full with letters, memos, reports, proposals, instructions, and other assignments. We rarely have time to teach a major research paper, nor do we believe that technical writing courses are the appropriate site for research papers. Summaries, however, lend themselves well to a technical writing course. They are a manageable length, and they can be taught along with other types of technical correspondence. Summaries are excellent assignments for teaching your students how to perform, use, and document research--without your having to teach a full blown research paper.

Here's how we often teach summaries. We present the following scenario to our students:

"You are a mid-level manager. Your boss, a very busy person, needs to give an oral presentation about technical innovations in your company's area of expertise. The speech will be given to a civic organization (Lion's Club, Rotary Club, etc.). Your boss, however, does not have the time to research this topic. He or she delegates the research to you. Your task is to research several articles in technical journals and then summarize the findings for the boss, providing him or her a capsulated version of the research. The boss then will read your summary to prepare for his or her speech."

With this scenario in mind, we do the following:

1. Give our students six to eight articles on technical innovations (hypertext, Internet, virtual reality, side impact protection devices for automobiles, hydroponic plant germination techniques, fire prevention safety measures, etc.).

2. Divide our class into small groups.

3. Ask the groups to read the articles and summarize them. This requires that they take marginal notes, organize their data, write a rough draft, and document their sources.

We then make transparencies of the summaries and review the collaborative efforts, making suggestions for revision. This simple activity simulates a real-world writing assignment, allows the students to practice teamwork, helps the students summarize, and asks them to document. A great deal is accomplished in a short time.

Activities

Using our end-of-chapter activities will give your students a better understanding of summaries.

Answers to Quiz Questions

1. A well-constructed summary contains an introduction with topic sentence, a discussion, and a conclusion.
2. The topic sentence will present the primary focus of the original source and list the two or three major points to be discussed.
3. You can organize a summary as the writer organized their information.
4. Omit irrelevant details, examples, explanations, or descriptions.
5. Your summary must be clear, concise, accurate, and accessible.

CHAPTER 16

REPORTS

- @ **Criteria for Writing Reports**
- @ **Types of Reports**
- @ **Process**

ABSTRACT

In this chapter, we explain the use of short reports in the work environment and provide you criteria for various types of short reports, including

- @ Trip Reports
- @ Progress Reports
- @ Lab Reports
- @ Feasibility Reports
- @ Incident Reports

We also provide *case* studies and team *projects.*

Teaching Suggestions

As we've noted before, collaborative writing projects offer students the best opportunity to practice technical writing skills. The same holds true for reports. Try collaborative writing activities to help your students better understand reports. Through team projects regarding reports, students not only learn to work together--an important concern--but they also can practice document design, clarity, conciseness, and audience recognition.

One group assignment that would work well is a progress report. Ask your students to write a report on their progress in your technical writing class. Their shared experiences will allow them to approach this assignment from a collective point of view.

In the introductory section of this progress report, they could state why they are taking your class (required for a degree program, elective, job-related, etc.) and what they had hoped to achieve. In the discussion section of the report, they could discuss the work they've accomplished and problems they've encountered (time restrictions, computer problems, group dynamics, etc.).

Finally, in the conclusion/recommendation, they could state what work they have left to complete and suggest ways in which you could improve the class or better meet their needs.

To accomplish this group activity, ask your students to prewrite using branching to gather their data, write a draft, and then revise their collaborative report. Once they've completed the assignment, you either could show the group reports on an overhead or have the students exchange their reports for peer review. In either case, the students will benefit from immediate feedback.

The case studies also provide your students an excellent opportunity to work collaboratively. In addition, rather than drawing from personal experience, as with the progress report, the case studies provide the students work-related scenarios for a team-written feasibility report. This requires greater objectivity on their part.

To help assess their success in this group activity, you could use the following Evaluation Checklist.

EVALUATION CHECKLIST
FEASIBILITY/RECOMMENDATION REPORT

1. Does the student (or group) provide a complete *Identification Line* (Date, To, From, Subject)?
   Yes
   No
2. Is the *Subject Line* capitalized and does it provide a focus + a topic?
   Yes
   No
3. Does the report's *Introduction* clearly state the purpose of the report and what problem motivated the report?
   Yes
   No
4. Does the report's *Discussion* explain what exactly the problem entails and/or how to solve this problem?
   Yes
   No
5. Does the writer include appropriate Graphics in the report's discussion unit to visually clarify the problem and/or solution?
   Yes
   No
6. Does the *Conclusion/Recommendation* sum up the situation and recommend the next step?
   Yes
   No
7. Is the report *Clear and Concise?*
   Yes
   No
8. Is the report *Accessible,* especially concerning a hierarchy of headings and subheadings?
   Yes
   No
9. Does the writer address *Audience* concerns by defining terms and involving the audience?
   Yes
   No
10. Has the writer minimized *Grammatical Errors?*
    Yes
    No

Activities

Following is a suggested revision for end-of-chapter activity #5.

DATE: February 17, 2003
TO: Sam Tintereto, Section Supervisor
FROM: Sylvia Light, Software Engineer
SUBJECT: MONTHLY PROGRESS REPORT ON DT1234 TEST EQUIPMENT PROJECT

Introduction
As part of this month's management by objectives (MBO) goal, I was responsible for working on Test Executives, Test Routines, and Peripherals. This is a report on my progress.

Discussion

Activity	*% Completion*
Test Executive	
-Entry	100
-Troubleshooting	90
-Calibration	90
-Enhancements	75
Test Routines	
-ProLog	25
-PreSet	0
-Flight	25
-Switch	25
-Contact	0
-Clear	0
Peripherals	
-6000	85
-5335	90
-1001	90
-4250	75

Test Routines are delayed due to a malfunction with our mainframe.

Conclusion/Recommendation
We are on schedule and should meet our March deadline. Currently, 75 percent of the project is completed. If you could provide access to a backup mainframe, we should avoid additional delays.

Answers to Quiz Questions

1. Reports supply a record of work, record and clarify complex information, present information to a large number of people, record problems encountered, document schedules, recommend future action, document current status, and record procedures.
2. Common types of reports are accident/incident reports, feasibility reports, inventory reports, staff utilization reports, progress reports, travel reports, lab reports, performance appraisal reports, study reports, and justification reports.
3. The four basic units of a report are heading, introduction, discussion and conclusion/recommendation.
4. The discussion of the report summarizes your activities and the problems you have encountered.
5. The reporter's questions can help you to organize the report and determine the content.
6. A Gantt chart clarifies status, showing where the writer is in relation to the completion of a task.
7. Branching allows you to depict information graphically so you can not only gather data but also visualize it.

CHAPTER 17

PROPOSALS

- @ Criteria for Proposals
- @ Process
- @ Sample Proposal

ABSTRACT

Proposals are more complex than short reports. Thus, they require more components. These include a title page, cover letter, table of contents, list of illustrations, abstract, introduction, body, conclusion, glossary, and related appendices. This chapter explains each of these components in detail and provides a complete proposal for your students' review.

Teaching Suggestions

To simplify the complexity of a proposal, we help our students visualize the components of their proposal by providing them the following graphic, a storyboard.

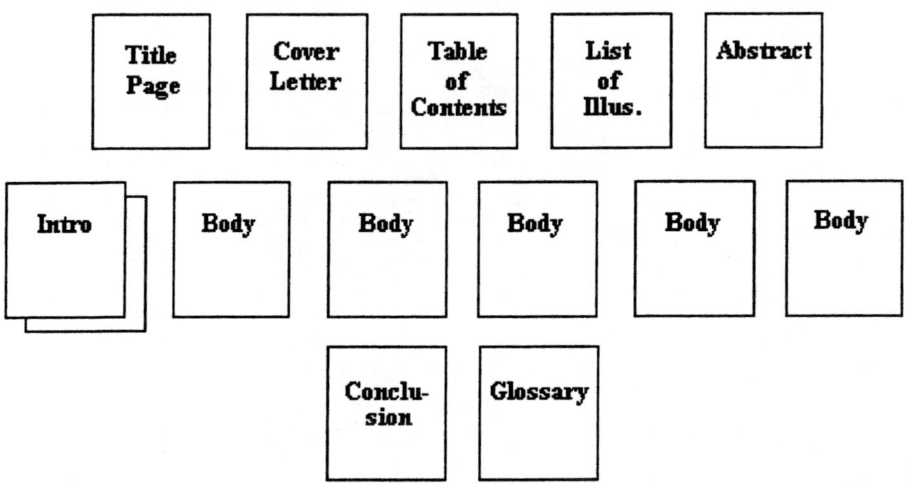

The above storyboard works well in conjunction with our FACT SHEETS, a valuable prewriting tool, which we've provided for you on the following pages. The FACT SHEETS will help your students organize their thoughts and gather data.

Finally, we always teach proposals as team projects. We divide students into groups. Their job then is to subdivide labors. They must manage their own project (the completion of the proposal) by designating who will write which of the above parts of the proposal, when each of these parts will be due for group review, who will complete the required graphics, etc.

Collaborative proposals help prepare students for the inevitable team-writing activities they will encounter on the job.

Fact Sheet

Abstract

1. Problem

   What is your audience's problem or area needing assistance? (Limit this statement to one to three sentences.)

2. Solution

   What is your suggested solution to this problem? (Limit this statement to one to three sentences.)

3. Benefit

   What benefit will your audience derive from your suggested solution? (Limit this statement to one to three sentences.)

Fact Sheet

Introduction

1. <u>Purpose</u>

   In one to three sentences, state the purpose of your long report.

2. <u>Problem</u>

   Elaborate on your audience's problem or area needing assistance. Doing so will show your knowledge of the situation and highlight the reader's need for your help. This should be precise, detailed, and quantified. Consider adding a graphic for clarity.

   - Problem #1

   - Problem #2

   - Problem #3

Fact Sheet

Main Text

Your report's main text could include any number of items. The following lists possibilities. Select any or all of them to develop your ideas.

- Analysis of solution (approaches)

- Technical Description

    ⇒ Mechanism

    ⇒ Tools

    ⇒ Equipment

    ⇒ Facilities

- Instructions

    ⇒ Maintenance

    ⇒ Repair

    ⇒ Installation

    ⇒ Enrollment

    ⇒ Display

    ⇒ Packing/Unpacking

    ⇒ Troubleshooting

- Options
    - ⇒ Approaches
    - ⇒ Personnel
    - ⇒ Products
    - ⇒ Facilities
    - ⇒ Financing

- Managerial Chains of Command

- Personnel Biographies

- Corporate and/or Employee Credentials
    - ⇒ Years in business
    - ⇒ Certifications
    - ⇒ Education/degrees
    - ⇒ Previous work history/accomplishments
    - ⇒ Satisfied customers

- Schedules
    - ⇒ Implementation
    - ⇒ Reporting

- ⇒ Maintenance
- ⇒ Delivery
- ⇒ Completion
- ⇒ Payment
- ⇒ Projected milestones (forecasts)

- Costs
  - ⇒ Base price/Unit price
  - ⇒ Quantity discounts
  - ⇒ Payment plans/financing

Fact Sheet

Conclusion/Recommendation

1. Conclusion

You could conclude your long report as follows:

- Restate audience's problem
- Restate your solution
- Restate the benefits derived
- Restate your credentials

2. Recommendation

You could recommend the next course of action. This could include

- When you plan a follow-up
- Reiteration of discounts offered before a certain deadline

Answers to Quiz Questions

1. An internal proposal is for your company's management.
2. An external proposal sells the benefits of a new corporate offering to a prospective client.
3. Use lowercase Roman numerals to number information prior to the introduction.
4. An abstract is a brief overview of the proposal's key points geared toward a low-tech reader.
5. The purpose statement informs your readers why you are writing and/or what you hope to achieve.
6. The discussion contains analyses, technical descriptions, instructions, options, managerial chain of command, biographical sketches of personnel, corporate and/or employee credentials, schedules, cost charts.
7. Your recommendation will suggest the next course of action. Specify when this action will or should occur and why that date is important.
8. Include a glossary because you might have readers with multiple levels of expertise.
9. Appendices allow you to include any additional information.
10. In a proposal you can use tables, figures and pictures.
11. Your writing style should be concise, precise and free of deadwood. Avoid old-fashioned words or words which are difficult to understand.
12. Format a proposal using headings and subheadings, itemized ideas, boldfaced key points, and underlined important words. Also use graphics and white space.
13. Enhance your tone by being positive and motivational.

CHAPTER 18

ORAL PRESENTATIONS

- @ **Types of Oral Presentations**
- @ **Criteria for Effective Oral Presentations**
- @ **Process**
- @ **Post-Speech Activities**

ABSTRACT

In this chapter, we discuss four types of speeches:

- @ The Impromptu Speech
- @ The Memorized Speech
- @ The Manuscript Speech
- @ The Extemporaneous Speech

In addition, we take your student from concept to finished product via the composing process--prewriting, writing (presentation plans and outlines), presentation, and post-speech activities.

Teaching Suggestions

We pair oral presentation with other types of technical correspondence. For example, after our students have written their long report or proposal, they then provide a five-to-ten minute briefing, highlighting the key components of their reports. This chapter will help your students prepare for such a briefing.

You also could use this chapter on oral presentations in conjunction with Chapters 9 and 7. Chapter 9 ("Graphics") will help your students combine oral presentation with visual aids. Oral presentations work especially well with Chapter 7: The Job Search). Have your students interview for a fictitious job, or one they've discovered in the classified ads. A group of students (or the entire class) can act as an interview committee.

After reviewing the letter of application and resume, the committee could interview the potential student applicant, whose job it would be to make a brief oral presentation on his or her skills. This would give your students valuable practice on a skill which will benefit them when they search for employment. And it's better to receive this training in the controlled environment of a classroom, with your assistance and tutoring, than on the firing line--when losing a job hurts.

Activities

Our end-of-chapter activities will suggest other ways to use oral presentation in your technical writing classes.

Answers to Quiz Questions

1. Oral presentations include the impromptu speech, the memorized speech, the manuscript speech, and the extemporaneous speech.
2. Introduce your presentation with an anecdote, facts and figures, a suitable joke, a question, a quotation, or a startling comment.
3. Organize your presentation by comparison/contrast, problem/solution, argument/refutation/persuasion, least to most important, and chronological.
4. A good presentation is clear, concise and recognizes audience.
5. Establish rapport with eye contact, rate of speaking, enunciation, pitch, pauses, and emphases.
6. You might be representing your company or trying to make a good impression for yourself when you speak, so dress appropriately.
7. In your presentation, use chalkboards, chalkless boards, flip charts, overhead transparencies, slides, videotapes, films, computerized presentations, and power point presentations.
8. Before giving your speech, consider if you are speaking to peers, to subordinates, to supervisors, to lay customers, to high tech or low tech people, to multiple levels of audience, and how well you know the audience.
9. Determine if your speech is trying to inform, persuade, explain procedures, or motivate.

UNIT 7

HANDBOOK

Chapter 19 Grammar, Punctuation, Mechanics, and Spelling

CHAPTER 19

GRAMMAR, PUNCTUATION, MECHANICS, AND SPELLING

@ **Grammar Rules**

@ **Punctuation**

@ **Mechanics**

@ **Spelling**

ABSTRACT

This chapter could be used as a mini-grammar handbook, providing your students rules for grammar, punctuation, mechanics, and spelling.

In addition, this chapter provides you and your students many abbreviations for commonly used technical terms and measurements, as well as rules for number usage.

Finally, the chapter offers numerous in-class and homework activities for further discussion.

Teaching Suggestions

You can use this chapter in one of two ways--either as a reference tool for your students' individual work at home or for in-class assignments and discussion.

Because most technical writing classes focus on numerous types of technical correspondence, little time is left for grammar lectures in class. If such is the case in your course, you could use this chapter only for reference or to help the student who has isolated grammatical needs. We have alphabetized the various grammatical errors so that your students can find these rules easily on their own.

In contrast, you might want to teach out of this chapter as an instructional unit within your technical writing course. That would be a great idea. Too many students think that grammar is an inconsequential nuisance--something that possibly lowers their grades but that has no real value in business and industry. They are mistaken in this notion. In fact, grammar is very important, not just to English teachers, but to anyone who reads the correspondence. Grammatical errors not only disrupt the reader's attention and confuse the reader, but errors also make the writer look bad. It's a lose-lose situation. We can emphasize the importance of grammar by devoting several weeks during a semester to teaching it to our students--not just as an afterthought but as a major component to the class.

This chapter is eminently teachable; we've tried it, and it works!

Activities

Spelling

Following is a list of the correctly spelled words from the sentences:

1. Except
2. Advice
3. Affect
4. Cite
5. It's
6. Lose
7. Quite
8. Their
9. Too
10. Your

Following is a list of the misspelled words from the letter and the words correctly spelled:

personel/personnel
advertizment/advertisement
Febuary/February
quallifications/qualifications
inclosed/enclosed
resum/resume
recieved/received
microprocesors/microprocessors
controll/control
ect./etc.
extrordinary/extraordinary
knowlege/knowledge
aquired/acquired
expirences/experiences
futher/further
foreward/forward
Sincerly/Sincerely

Fragments and Comma Splices

Following are suggested corrections for the sentences.

1. She kept her appointment with the salesperson.
2. When the CEO presented his fiscal year projections, he tried to motivate his employees. Many were not excited about the proposed cuts.
3. The company's sales were up 25 percent.
4. The supervisor wanted the staff members to suggest improvements for their work environment. The employees, however, felt that any grievances should be taken directly to their union representatives.
5. He decided it was an excellent idea.
6. Their machinery was prone to malfunctions and often caused hazards to the workers.
7. They needed the equipment to complete their job responsibilities. Further delays would cause production slowdowns.
8. Their client was a major distributor of high-tech machinery.

Following are suggested corrections for the errors in the letter. We've corrected the errors in the order they appear in the letter. Numerous options exist:

- The purpose of the report was to determine the effectiveness of the system and to offer any concrete recommendations for improvement.

- To do this, I analyzed in detail the damage cost figures for the past three years. I also did an extensive study of dormitory conditions.

- Although I had limited human resources, I gathered information on all seven dormitories, focusing specifically on the men's athletic dorm, located at 1201 Chester.

- In this dorm, bathroom facilities, carpeting, closet doors, and air conditioning are most susceptible to damage.

- Nonetheless, my immediate findings indicate that the system is functioning well. However, improvements in the physical characteristics of the dormitories, such as new carpeting and paint, would make the system even more efficient.

- I have enjoyed conducting this study. I hope my findings help you make your final decision.

- Please contact me if I can be of further assistance.

Punctuation

Following is a list of the needed punctuation in the sentences:

1. (,)
2. (;)
3. (;)
4. (,) (/)
5. (:) (;) (;)
6. (;)
7. (/)
8. (,)
9. (,) (/)
10. (/) (/) (;) (,)

Following are corrections to the letter, provided in order of appearance:

- 8,
- Mr.
- Carbondale, IL
- Mr. Schaefer:
- Yesterday, my partners and
- Tribune, and
- idea and
- project.
- project.
- However,
- so,
- questions:
- water?
- part?
- consulting?
- Once I
- questions, my
- operations.
- concept and
- tremendously.
- you.
- Sincerely,

Agreement (Subject/Verb and Pronoun/Antecedent)

Following are corrections to the sentences:

1. Were
2. He
3. Was
4. His or her
5. Work
6. Works
7. He
8. He or she
9. Are
10. Has

Following are corrections to the memo, provided in order of appearance:

- There have been several incidents involving the unit which have resulted . . .
- The occurrences yesterday were caused . . .
- To prevent recurrence of this type of damage, the following actions
- have been . . .
- Each supervisor must insure that his or her . . .
- Everyone must be made aware that he or she is
- The supply air duct, modified to carry away harmful sediments, is . . .
- Maintenance supervision recommends . . .

Following are corrections to the memo:

DATE: December 5, 2003
TO: Jordan Cottrell
FROM: Richard Davis
SUBJECT: SELF-CONTAINED BREATHING APPARATUS (SCBA) AND NEGATIVE RESPIRATOR EVALUATION AND FIT-TESTING REPORT

The evaluation and fit-testing have been accomplished. The attached list identifies the following:

- Supervisors and electronic technicians who have used the SCBA successfully.
- The negative pressure respirators used for testing in an isoamle acetate atmosphere.

Fit-testing of waste management personnel will be accomplished annually, according to President Chuck Carlson. New supervisors and technicians will be fit-tested when hired.

All Apex Corporation personnel located in the New York district must submit a request form when requesting a respirator or SCBA. Any waste management personnel in the north and south facilities not identified on the attached list will be fit-tested only when they plan to use an SCBA. If you have any questions, contact President Carlson or me (Richard Davis) at ext. 4036.

Answer Key to Grammar Quiz

1. A
2. C
3. C
4. B
5. A
6. D
7. A
8. A
9. B
10. A
11. B
12. C
13. C
14. B
15. B
16. B
17. A
18. A
19. B
20. C
21. D
22. A
23. C
24. D
25. A

EXAMS

Exam 1: Chapters 1-4
- An Introduction to Technical Writing
- Producing the Product
- Objectives in Technical Writing
- Audience Recognition and Involvement

Exam 2: Chapters 5-7
- Memos
- Letters
- The Job Search

Exam 3: Chapters 8-9
- Document Design
- Graphics

Exam 4: Chapter 10
- Flyers, Brochures, and Newsletters

Exam 5: Chapters 11-13
- Technical Descriptions
- Instructions and User's Manuals
- E-mail, Online Help, and Web Sites

Exam 6: Chapters 14-18
- Research
- The Summary
- Reports
- Proposals
- Oral Presentations

Exam 1: Chapters 1-4

1. Technical and business writing are considered writing of which type?
    a. creative
    b. expository
    c. expressive
    d. persuasive
    e. none of the above

2. The purpose of technical/business writing is often determined by audience (T/F).

3. The textbook, in chapter one, offers four different reasons why this type of writing is important. One of these reasons is that
    a. your purpose is determined by your audience
    b. technical and business writing differ from other types of writing
    c. this type of writing costs money
    d. technical and business writing depend on factual documentation

4. The writing process is sub-divided into the following three-part sequence:
    a. prewriting, revising, rewriting
    b. prewriting, drafting, proofreading
    c. prewriting, writing, rewriting
    d. prewriting, rewriting, revising
    e. none of the above

5. Prewriting techniques are helpful, but they won't assist you in overcoming "Blank Page Syndrome" (T/F).

6. Two examples of prewriting techniques:

    a. mindmapping and brainstorming
    b. simplifying and reformatting
    c. flowcharting and reformatting
    d. branching and enhancing
    e. none of the above

7. Which of the following is a guideline for establishing teamwork?
    a. establish a team leader
    b. avoid shun words
    c. watch for cultural biases
    d. hold yourself responsible for how well your audience understands your message

8. To produce successful technical/business writing, you need to approach writing as a process (T/F).

9. High-tech readers work in your field of expertise (T/F).

10. One way to avoid confusion when using acronyms and abbreviations is to define them parenthetically or in a glossary (T/F).

11. Which of the following is a passive-voice sentence?

    a. The salesman will probably close the deal tomorrow.
    b. He will close the deal tomorrow.
    c. The deal will be closed tomorrow.
    d. Truthfully, I hope he closes the deal tomorrow.

12. Which of the following contains no passive-voice structures?
    a. While the room was being cleaned by the maid, we waited outside.
    b. We couldn't see the sky because the shades had been closed and the curtains had been drawn by whomever was in the room last.
    c. How old will Steve be on his next birthday?
    d. The dog was shot by an irate neighbor.

13. The text lists six methods of improving conciseness. Which one of the following is not one of them?
    a. limit paragraph length
    b. use the meat cleaver theory of revision
    c. avoid wordy phrases
    d. use the gestation approach

14. The ultimate goal of good technical/business writing is clarity (T/F).

15. One way to achieve good organization is by supplying specific, quantified information (T/F).

16. Accuracy in technical writing requires that you proofread your text (T/F).

17. Your text lists ten accuracy-improving tips. Two of them are

    a. let someone else read it and avoid shun words
    b. read it aloud and try scattershot proofing
    c. read it aloud and omit redundancies
    d. read one line at a time and use emphatic words
    e. none of the above

18. To compose effective technical/business writing, you must achieve audience recognition and involvement (T/F).

19. Writing to only high-tech audiences is relatively easy because you can use acronyms, abbreviations, jargon, complex graphics, etc. (T/F).

20. The five primary goals of effective technical/business writing are clarity, conciseness, accuracy, organization, and ethics (T/F).

21. Customers and clients who neither work for your company nor have any knowledge of your field of expertise are considered a low-tech audience (T/F).

22. Strategies for successful collaboration mentioned by your textbook authors do not include

    a. choose a team leader
    b. analyze the problem's cause
    c. provide specific detail
    d. verify the suggested solutions
    e. none of the above

23. The disadvantages of teamwork are outweighed by its benefits (T/F).

24. A document you write should be legal, practical, and ethical, which are three separate and unrelated issues (T/F).

25. Which of the following is a guideline for establishing ethical standards?
    a. avoid shun words
    b. establish a team leader early in the project
    c. watch for cultural biases
    d. hold yourself responsible for how well your audience understands your message

Answer Key to Exam 1: Chapters 1-4

1. E
2. T
3. C
4. C
5. F
6. A
7. A
8. T
9. T
10. T
11. C
12. C
13. D
14. T
15. T
16. T
17. B and D
18. T
19. T
20. T
21. F
22. C
23. T
24. F
25. A

Exam 2: Chapters 5-7

1. Memos and email are written more than any other type of technical correspondence (T/F).

2. Your audience for a memo you write will probably be more "high-tech" than your audience for a letter (T/F).

3. Your textbook claims that no one method of prewriting is the most effective (T/F).

4. Memos are important not only because of their frequency of use and wide range of subject matter but also because they represent a major component of your interpersonal skills within your work environment (T/F).

5. Unlike letters, writing errors in memos won't affect the writer's credibility (T/F).

In items 6 through 15, each of the characteristics applies to either letters or memos. If it applies to letters, mark "A"; if it applies to memos, mark "B."

6. Informal tone

7. High-tech subject matter with acronyms and abbreviations often allowed

8. External correspondence

9. Always includes an inside address

10. Generally, no standard formats; depends on how your company does it.

11. "Enclosure" notation at the bottom of the page

12. Formats include "full-block" and "indented"

13. Usually includes a salutation

14. Always includes a subject line

15. Requires adjustment of margins, depending on the length of the message

16. The first paragraph of an effective letter should tell your reader what you want to happen next,when you want this to happen, and why it is important (T/F).

17. One of the innate problems with email is that it does not offer complete documentation in the way that a letter or memo does (T/F).

18. Unlike memos, in email using all capital letters is acceptable (T/F).

19. The internet is a primary conduit for technical information (T/F).

20. Your textbook authors state that you should never use "Miss" and "Mrs." in the salutation of a letter (T/F).

Write your responses to the following items on the back of the Scantron form. Be sure to number your responses clearly.

21. What are the two types of conclusions or closes in a memo? (This question is not referring to the closes used in letters, like "sincerely.")

22. Which type of letter is used to focus your reader's attention and tell your reader about what he or she is receiving when you are sending invoices, drawings, copies of reports, questionnaires, proposals, your resume, etc.?

23. Two types of letters begin with buffer statements in the first paragraph before you present your decision in the discussion section of the letter. Name one of them.

24. The discussion section of this type of letter persuasively explains why your reader should purchase your product or service.

25. A letter you write in which you offer someone a job or inform someone of a promotion or an award is an example of what type of letter?

26. In the discussion of which type of letter do you specify your needs?

27. If your letter is longer than one page, what notation do you place at the top of the additional page(s)? Be specific.

28. Very specifically, what are the two formats your textbook says you can use for this information mentioned in item 27 above? Where, that is, and in what arrangement would these notations be found at the top of the second page of a letter?

29. One of the best ways to find a job is to search on the internet (T/F).

30. You will never find a job by networking with friends or past employers (T/F).

31. Do not write a reverse chronological resume if you
    a. plan to leave your present professions
    b. are a traditional job applicant
    c. hope to enter the profession in which you earned your college degree
    d. have made steady progress in one profession

32. Omit one of the following in your resume:
    a. identification
    b. career objectives
    c. religion
    d. employment history
    e. education

33. You should write a generic career objective, the more vague the better (T/F).

34. Be sure to have large gaps of unaccounted for time in your employment and educational history (T/F).

35. The incorrect style of writing in a resume includes
    a. itemization
    b. long sentences
    c. reader-friendly format
    d. positive, active verbs

36. Some benefits of online resumes are all but one of the following:
    a. allow thousands of international employers to access it instantly
    b. allow you to attach lengthy documents, full-color graphics, animation and sound
    c. allow you to seek help from potential employers particularly with your software use
    d. allow you to update the resume quickly and easily

37. State the software you're using and the version when you create an online resume (T/F).

38. You can use highlighting techniques such as italics, underlining, and color in your online resume (T/F).

39. You should omit in the introduction to your letter of application
    a. where you discovered the job opening
    b. desired salary
    c. specific job for which you're applying
    d. your best credential

40. In the body to a letter of application, be sure to include how the job will make you happy (T/F).

41. Never include in the conclusion of the letter of application when you will get in touch with the business (T/F).

42. When you interview, you should do all but one of the following:
    a. wear heavy jewelry
    b. avoid chewing gum, smoking, or drinking coffee
    c. watch your body language
    e. control your speaking

43. Online job sites, such as Monster.com, let you post your resume online, provide information about job opportunities globally as well as locally, and share tips for writing effective resume. (T/F)

Answer Key to Exam 2: Chapters 5-7

1. T
2. T
3. T
4. T
5. F
6. B
7. B
8. A
9. A
10. B
11. A
12. A
13. A
14. B
15. A
16. F
17. T
18. F
19. T
20. F
21. Directive or Complimentary
22. Cover letters
23. Bad news letters
24. Sales letters
25. Good news letters
26. Letter of complaint or inquiry
27. Your name, page number, and the date
28. Left margin, subsequent page notation
    Name
    Page number
    Date

    Or across the top of subsequent pages
    Name page number date
29. T
30. F
31. A
32. C
33. F
34. F
35. B
36. C
37. T

38. F
39. B
40. F
41. F
42. A
43. T

Exam 3
Open book

Exam 3: Chapters 8-9

1. Visually unappealing and inaccessible correspondence can negatively affect your company's sales and reputation (T/F).

2. Achieving effective document design is dependent on all but one of the following visual appeals?
   a. organization
   b. order
   c. access
   d. entertainment
   e. variety

3. Breaking text into smaller chunks of information, chunking, is an ineffective way to communicate (T/F).

4. You will achieve chunking with all but one of the following techniques:
   a. story boarding
   b. headings
   c. white space
   d. rules
   e. section dividers and tabs

5. Establishing a hierarchy of importance in your headings does not contribute to accessibility (T/F).

6. To help readers filter out extraneous or tangential information and focus on key ideas, you should not do one of the following:
   a. add white space
   b. use reverse chronology
   c. create lists
   d. use bullets
   e. enclose key ideas in a box

7. It's impossible to overuse highlighting techniques (T/F).

8. You will not achieve variety of document design if you do one of the following:
   a. by adding longer paragraphs
   b. by printing horizontally
   c. by using more columns
   d. by varying gutter width
   e. by using ragged-right margins

9. With effective document design, you can save money and time (T/F).

10. Use of visual aids will not help you achieve
    a. subjective writing
    b. conciseness
    c. clarity
    d. cosmetic appeal

11. Graphics assist readers in seeing all but one of the following:
    a. trends
    b. comparisons between like components
    c. style
    d. percentages
    e. facts and figures

12. Because a color rarely goes out of style, you should use one which is in fashion today (T/F).

13. You can specify and quantify with a 3-D graphic (T/F).

14. Rules for effective use of graphics do not include one of the following:
    a. Don't integrate a graphic with text.
    b. Appropriately locate graphics.
    c. Use a graphic to add to text.
    d. Use a graphic to communicate important information that could not be conveyed easily with a paragraph.

15. Tables should be numbered in order of presentation (T/F).

16. You don't need to use text to lead into a table (T/F).

17. Consistency in numbering style in a table is necessary (T/F).

18. Whereas tables eliminate needless repetition of words, figures highlight and supplement important points in your writing (T/F).

19. A Gantt chart depicts a chronological history of an activity (T/F).

20. You should use at least 15 or 20 wedges in a pie chart (T/F).

21. Icons are visual representations of a capability, a danger, a direction, an acceptable behavior, or an unacceptable behavior (T/F).

22. Icons should not have one of the following characteristics:
    a. simplicity
    b. realism
    c. cultural and gender stereotyping
    d. recognizability
    e. universality

23. You can download any image from any website because these images are freeware (T/F).

24. Freeware means you can download for a very small fee or risk infringement on copyright laws (T/F).

25. It's impossible to modify downloaded graphics (T/F).

Answer Key to Exam 3: Ch. 8-9

1. T
2. D
3. F
4. A
5. F
6. B
7. F
8. A
9. T
10. A
11. C
12. F
13. F
14. A
15. T
16. F
17. T
18. T
19. T
20. F
21. T
22. C
23. ~~T~~ F - either T or F. Can be interpreted either way
24. F
25. F

Exam 4: Chapter 10

1. Which of the following is not true about effective brochures?
   a. cost effective
   b. time efficient
   c. many pages long
   d. responsive to immediate needs
   e. personalized

2. If you keep your flyer to one page (front only), you can use the back for mailing information (T/F).

3. A flyer should contain many topics and discuss them in great detail (T/F).

4. To create your flyer, you should avoid doing one of the following:
   a. use a title
   b. have unlimited text
   c. increase font size
   d. use graphics
   e. use color

5. It's impossible to overdo highlighting techniques in a flyer (T/F).

6. Using a catchy phrase is a good idea for your flyer (T/F).

7. Brochures differ from flyers because brochures are much shorter than flyers (T/F).

8. A typical title page in a brochure omits
   a. the product or service name
   b. a visual representation of the topic
   c. prices
   d. contact information

9. On the back panel of the brochure, consider excluding
   a. highlights of the topic
   b. mailing information
   c. coupons
   d. detailed analysis of uses of the product or service
   e. location

10. Although headings and subheadings are helpful navigational tools to guide your reader, you should not use them in a brochure because they take up space (T/F).

11. Designing your brochure entails all but one of the following?
    a. limit sentence length to 20 or 30 words
    b. use white space
    c. use color
    d. bulletize key points
    e. boldface and/or underline key ideas

12. You should avoid alternating placement of visuals in a brochure (T/F).

13. A personal approach is best in a brochure, so you should use pronouns and contractions (T/F).

14. Cubing is a prewriting technique that allows you to view a topic from six perspectives (T/F).

15. To determine the usability of your brochure, you should not consider
    a. retrievability
    b. readability
    c. accuracy
    d. audience analysis
    e. user satisfaction

16. Newsletters allow you to share detailed information in a personal and friendly manner (T/F).

17. Parts of a newsletter do not include
    a. the flag
    b. columns
    c. an index
    d. headings and subheadings
    e. pop-up quotes

18. Avoiding lengthy paragraphs is important in your newsletter (T/F).

19. It's possible to find flyers, brochures, and newsletters on the internet (T/F).

20. Picking four or five dominant colors is a good idea when creating a flyer (T/F).

21. Several of the details to include in a brochure are technical specifications, company history, credentials, delivery dates, and payment plans (T/F).

22. You should avoid allowing text and graphics to overlap two or more panels in a brochure (T/F).

23. Avoid definitions of technical terms or abbreviations in a brochure (T/F).

24. Usability is a way of determining whether your reader can use a brochure effectively (T/F).

25. Most often, newsletters are 8 ½ " X 11" with a portrait orientation in five to six columns (T/F).

Answer Key to Exam 4: Chapter 10

1. C
2. T
3. F
4. B
5. F
6. T
7. F
8. C
9. D
10. F
11. A
12. F
13. T
14. T
15. D
16. T
17. C
18. T
19. T
20. F
21. T
22. F
23. F
24. T
25. F

Exam 5: Chapters 11-13

1. Technical descriptions are important features of
    a. operations manuals
    b. product demand specifications
    c. study reports provided by consulting firms
    d. sales literature and proposals
    e. all of the above

2. To arrange a technical description "spatially" (as opposed to chronologically, geographically, in order of importance, etc.), means organizing the components as they exist in space. (T/F)

3. "Impressionistic" word usage and "photographic" word usage are essentially the same thing. (T/F)

4. Technical descriptions are a part-by-part depiction of how the components of a mechanism, tool, or piece of equipment are to be operated by the user. (T/F)

5. The conclusion of your technical description depends on the purpose for which you are preparing the document. (T/F)

6. Possible types of conclusions for technical descriptions include repetition of introductory comments, comparison/contrasts to competitor's merchandise, testimonies, and guarantees. (T/F)

7. To describe the topic of your technical description clearly and accurately, you will often need to detail
    a. weight
    b. materials
    c. dimensions
    d. identifying numbers
    e. all of the above

8. Graphics that are appropriate for a technical description include
    a. bar charts and pictographs
    b. pie charts and Gantt charts
    c. exploded views and sectional cutaways
    d. line graphs and flow charts
    e. all of the above

9. A company sometimes issues a "proposal" when it needs a piece of equipment that does not currently exist. (T/F)

10. An example of a type of document in which technical descriptions are rarely if ever seen is sales literature. (T/F)

11. You should never use sexist language in technical description. (T/F)

12. The names or numbers that indicate or label particular parts of a line drawing or photograph are called _____________.

13. Using figures drawn in cartoon-like characters to highlight parts of a graphic and to interest readers is called the _____________.

14. Sometimes a company needs a piece of equipment that does not currently exist. To acquire this equipment, the company writes a _________________ specifying its exact needs.

15. The three-part writing process used in producing a summary or a technical description:

    ____
    ____
    ____

16. Whenever a company manufactures a product, it encloses in the packaging an _________________ that helps the end user construct, install, operate, or service the equipment.

17. What is the document called that is produced by a consulting engineering firm that has been hired to study a problem and provide a descriptive analysis?

18. A neatly organized set of instructions can orient readers, thereby helping them follow your directions (T/F).

19. As well as helping your readers, graphics in your instructions also make the document more visually appealing (T/F).

20. You are more likely to use graphics in a set of instructions for a high-tech end-user than a low-tech end-user (T/F).

21. Always use bullets for each step (T/F).

22. Assuming you are writing a step correctly, it will begin with a verb (T/F).

23. Limit the information within each step by separating distinct, small items into their own steps (T/F).

24. Highlighting techniques are particularly important
    a. in the conclusion
    b. in the introduction
    c. when presenting warnings and cautions
    d. if the writer will also be the end user
    e. all of the above

25. For writing instructions, asking reporter's questions is a good method to start gathering data and determining objectives. (T/F)

26. Using pronouns in instructions
    a. is something you should generally avoid
    b. is one way to keep your instructions concise
    c. helps add a person-to-person touch
    d. is more effective than using graphics

27. Unlike other types of writing, sexist language is acceptable in instructions. (T/F)

28. One of the biggest problems encountered when writing instructions is deciding how much detail to include. (T/F)

29. When writing instructions, what you might see as unnecessary, your reader might see as necessary. (T/F)

30. When including hazard alert messages in instructions, you need to consider colors, text, icons, and which of the following three things:
    a. placement, access, and conclusions
    b. placement, access, and definitions
    c. access, definitions, and checklists
    d. access, definitions, and specifications
    e. none of the above

31. What your reader has difficulty understanding, or you have difficulty writing clearly, your graphics can help explain pictorially in a set of instructions. (T/F)

32. The key to success as a writer of instructions is "Don't assume anything. Spell it all out—clearly and thoroughly!" (T/F)

Please write your responses to the following questions on the back of the Scantron form and number them clearly.

33. The textbook refers to drawings that are big, simple, and clear as ________________ graphics and recommends that you use them in instructions.

34. Instructions—such as the statement "Turn the knob counter-clockwise"—are always written in which *mood*?

35. Steps in sets of instructions are arranged in what order?

36. Give three examples of the words that "reporter's questions" begin with.

37. Give an example of an item that your textbook suggests might be included in the conclusion of a set of instructions.

38. Sequencing your steps correctly is best accomplished through a pre-writing technique called ____________________.

39. What does ANSI stand for?

40. The textbook suggests a hierarchy of definitions for hazard alert messages (the four words generally used in instruction writing to indicate potentially harmful or hazardous situations). Which of these four words does the textbook suggest that you use for the definition "The potential for death"?

41. The textbook makes the distinction between a short set of instructions and a *user's manual* by discussing nine different components that might be found in a manual but not found in a short set of instructions. Name *three* of them. Be specific.

42. Your textbook suggests that you do not omit "articles" when writing instructions. Give two examples of "articles."

43. The Help menus on your computer are excellent examples of __________ text online documentation. This "exciting new aspect" of technical communication links bits and pieces of information together electronically, allowing the writer to create interactive training tools within a document. It's also used extensively for "links" and movements from one page or site to another on the World Wide Web.

45. Your text gives eight steps for the rewriting stage of web-site construction. What is one of them?

46. Your text gives three reasons why on-line help systems are growing in importance. What is one of them?

46, 47, and 48. What are three of the seven suggestions your text provides for writing effective on-line help?

____

____

____

Answer Key to Exam 5: Chapters 11-13

1. E
2. T
3. F
4. T
5. T
6. T
7. E
8. C
9. F
10. T
11. T
12. callouts
13. super comic book look
14. product demand
15. prewriting, writing, and rewriting
16. operations manual
17. study report
18. T
19. T
20. F
21. F
22. T
23. T
24. E
25. T
26. A
27. F
28. T
29. T
30. B
31. T
32. T
33. super comic book
34. imperative
35. chronological
36. who, what, where, when, why
37. a warranty
38. flowcharting
39. American National Standards Institute
40. danger
41. Table of Contents, Introduction, Hazard Alerts, Procedural Steps
42. A, an, the
43. hypertext
44. Add details, delete unnecessary information, simplify words and phrases

45. Increased use of computers in business, industry, education, and home
46. Know your audience
47. Identify yourself
48. Keep your e-mail message brief

Exam 6: Chapters 14-18

1. Rules of documentation are structured rigidly to avoid plagiarism and to create uniformity (T/F).

2. The most effective style of writing in a research paper is subjective (T/F).

3. When you write a research paper, consider all but one of the following:
    a. audience
    b. purpose
    c. subjective interpretation of facts
    d. facts

4. Using the internet to research a topic should be avoided (T/F).

5. Two types of outlines include the topic outline and the paragraph outline (T/F).

6. A general rule for using source material in your research paper is to do all but one of the following:
    a. make a statement before the quotation or paraphrase
    b. support this generalization with a quotation or paraphrase
    c. provide a follow-up explanation
    d. use only one quotation or paraphrase in any one paragraph

7. Depending on discipline for which you are writing the research paper, you can rely on either the Modern Language Association or the American Psychological Association style sheet to document your sources (T/F).

8. The body of a summary should follow the organization of the original article (T/F).

9. To develop a summary of an article, you need to focus on the article's
    a. most important points
    b. major conclusions
    c. recommendations
    d. all of the above

10. The types of things found in the original article that should be omitted in a summary are
    a. key ideas and significant points
    b. recommendations and conclusions
    c. findings and conclusions
    d. biographical information and statistics
    e. all of the above

11. Information that is interesting but not mandatory for your reader's understanding should be left out of a summary (T/F).

12. A summary should not present your own ideas (T/F).

13. Using sexist language is permissible in summaries (T/F).

14. A summary is like a Reader's Digest approach to writing: although the summary will not cover every fact in the original, after reading the summary the reader should have a clear overview of the original's main points (T/F).

15. A summary should be ______ to ______ percent of the length of the original.

16. No basic traits are common to all reports. (T/F)

17. All reports share certain similarities in format, development, and style. (T/F)

18. Some writers skip the introductory portion of reports and go directly to the discussion section. This is OK if they believe that their readers know why the reports were written and who is involved. (T/F)

19. The discussion section is usually the largest section of the report. (T/F)

20. A list of illustrations is not needed in a proposal if it already has a table of contents. (T/F)

21. You should use highlighting techniques in a report. (T/F)

22. The purpose section of a report is usually one of the longest sections. (T/F)

23. Unfamiliar terms should be defined in the body of a proposal, not in a glossary. (T/F)

24. An appendix is a good place to file data that the reader may not immediately need but which provides documentation for further reference. (T/F)

25. A purpose statement is like a topic sentence. (T/F)

26. A purpose statement explains why you are submitting the report and what the report's subject matter is. (T/F)

27. Graphics and charts used in reports should be placed in an appendix. (T/F)

28. Sexist language is permissible in reports. (T/F)

29. Incident reports are written at regularly scheduled intervals. (T/F)

30. The organization of a report should
    a. move chronologically
    b. begin with generalizations and then move through specific details
    c. begin with details and move toward generalities
    d. begin with details and move toward a conclusion
    e. all of the above

31. Reports are written in a subjective style. (T/F)

32. The cover memo or cover letter of a report
    a. is rarely used today in modern reports
    b. is composed of a purpose, discussion, and recommendation
    c. is numbered as page I in the table of contents
    d. provides the function of "transmitting" the report from the writer to the reader
    e. all of the above

Write your responses to the following questions on the back of the Scantron form. Be sure that you number your responses.

Label the following five items according to what type of report they would appear in: trip, progress, lab, feasibility, or incident:

33. work actually accomplished to date

34. the steps you followed in performing the test

35. a review of your travel and activities

36. costs, specifications, labor requirements, availability, potential problems, etc.

37. apparatus used

38. What are the three steps in the writing process that your textbook authors suggest you follow when creating a report?

39 and 40. What are two of the four basic units in a report?

41. What type content do you include in the body of a trip report?

42. The ____________________section of a report constitutes the body of a report.

43. Unfamiliar terms in a report are often defined at the end of the report in a section called a _________________________.

44. Additional sections of information that are not appropriate for the body of the report but need to be included for the reader's reference and are added to the end of the report are called an ____________________.

45. ______________________allows you to depict information graphically.

46. If you're using research to write your proposal, you will need to include a __________________page.

47. Another name for "incident" report is ____________________________report.

48. A report that answers the question "Is labor sufficient and efficiently used" is referred to as a ___________________________report.

49. A report that answers the question "How is this particular employee doing on the job?" is referred to as a _________________________report.

50. Which of the following is not a type of oral presentation
    a. the impromptu speech
    b. the concluding speech
    c. the memorized speech
    d. the manuscript speech

51. The best type of speech to present is the memorized speech (T/F).

52. Ways to catch reader attention in a speech include an anecdote, startling comment, and facts or figures (T/F).

53. In a speech, once you've made a point, never repeat this same point (T/F).

Answer Key to Exam 6: Chapters 14-18

1. T
2. F
3. C
4. F
5. F
6. D
7. T
8. T
9. D
10. D
11. T
12. T
13. F
14. T
15. 5 to 15%
16. F
17. T
18. F
19. T
20. T
21. T
22. F
23. F
24. T
25. T
26. T
27. F
28. F
29. F
30. E
31. F
32. D
33. progress
34. lab
35. trip
36. feasibility
37. lab
38. prewriting, writing, rewriting
39. and 40. heading and introduction
41. findings
42. discussion
43. glossary
44. appendix
45. branching

46. works cited
47. trouble report
48. feasibility
49. progress
50. B
51. F
52. T
53. F